Mathematical Games

Five other collections of Martin Gardner's columns from *Scientific American* have been published as books (by Simon and Schuster): They are:

The *Scientific American* Book of Mathematical Puzzles and Diversions (1959)

The Second *Scientific American* Book of Mathematical Puzzles and Diversions (1961)

New Mathematical Diversions from *Scientific American* (1966)

The Numerology of Dr. Matrix (1967)

The Unexpected Hanging and Other Mathematical Diversions (1968)

Martin Gardner's
Sixth Book of
Mathematical Games
from
Scientific American

W. H. FREEMAN AND COMPANY
San Francisco

Library of Congress Catalogue Card Number: 75–157436
Standard Book Number: 0–7167–0944–9

Printed in the United States of America

9 8 7 6 5 4 3 2 1

For my brother-in-law, James B. Weaver

Contents

Introduction

Ten years ago the writer of a mathematics textbook would have been considered frivolous by his colleagues if his book included puzzles and other entertaining topics. This is no longer true. Exercises in the first two volumes of Donald E. Knuth's monumental work in progress, *The Art of Computer Programming* (New York: Addison-Wesley, 1968, 1969), are filled with recreational material. There are even textbooks in which a recreational emphasis is primary. A delightful instance is Harold R. Jacobs's *Mathematics: A Human Endeavor,* subtitled *A Textbook for Those Who Think They Don't Like the Subject* (San Francisco: W. H. Freeman and Co., 1970). Richard Bellman, Kenneth L. Cooke, and Jo Ann Lockett, authors of *Algorithms, Graphs, and Computers* (New York: Academic Press, 1970), write in their preface, "The principal medium we have chosen to achieve our goals is the mathematical puzzle."

The trend is not hard to understand. It is part of the painfully slow recognition by educators that students learn best who are motivated best. Mathematics has never been a dreary topic, although too often it has been taught in the dreariest possible way. There is no better way to relieve the tedium than by injecting recreational topics into a course, topics strongly tinged with elements of play, humor, beauty, and surprise. The greatest mathematicians always looked upon their subject as a source of intense intellectual delight and seldom hesitated to pursue problems of a recreational nature. If you flip the leaves of W. W. Rouse Ball's classic British work, *Mathematical Recreations and Essays* (first published by Macmillan in 1892 and soon to be issued in a twelfth revised edition), you will find the names of celebrated mathematicians on almost every page.

Euclid himself, among the earliest of the mathematical giants, wrote an entire book (unfortunately it did not survive) on geometrical fallacies. This is a topic covered in standard works on recreational mathematics but curiously avoided in most geometry textbooks. One of these days high school teachers of geometry will discover that an excellent way to impress their students with the need for rigor in deduction is to "prove" on the blackboard that, say, a right angle equals an obtuse angle, then challenge the class to explain where the reasoning went wrong.

The value of recreational mathematics is not limited to pedagogy. There are endless historical examples of puzzles, believed to be utterly trivial, the solving of which led to significant new theorems, often with useful applications. I cite only one recent instance. Edward F. Moore writes, in an important paper on "The Shortest Path through a Maze": "The origin of the present methods provides an interesting illustration of the value of basic research on puzzles and games. Although such research is often frowned upon as being frivolous, it seems plausible that these algorithms might eventually lead to savings of very large sums of money by permitting more efficient use of congested transportation or com-munication systems." (Reprinted in *Annals of the Computation Laboratory of Harvard University*, Vol. 30, 1959; pages 285–292.) Need I remind readers that the maze is a topological puzzle older than Euclid's geometry, and that topology itself had its origin in Leonhard Euler's famous analysis of a route-tracing puzzle involving the seven bridges of Königsberg?

This is the sixth anthology of my articles for the *Scientific American* department called Mathematical Games. As in previous collections, the articles have been expanded, errors corrected, bibliographies added. I am grateful to the magazine for the great privilege of contributing regularly to its pages, to my wife for unfailing help in proofing, and as always to the hundreds of *Scientific American* readers whose suggestions have added so much to the value of the original articles.

MARTIN GARDNER

February, 1971

1. The Helix

Rosy's instant acceptance of our model at first amazed me. I had feared that her sharp, stubborn mind, caught in her self-made antihelical trap, might dig up irrelevant results that would foster uncertainty about the correctness of the double helix. Nonetheless, like almost everyone else, she saw the appeal of the base pairs and accepted the fact that the structure was too pretty not to be true.

James D. Watson, *The Double Helix*

A STRAIGHT SWORD will fit snugly into a straight scabbard. The same is true of a sword that curves in the arc of a circle: it can be plunged smoothly into a scabbard of the same curvature. Mathematicians sometimes describe this property of straight lines and circles by calling them "self-congruent" curves; any segment of such a curve can be slid along the curve, from one end to the other, and it will always "fit."

Is it possible to design a sword and its scabbard that are *not* either straight or curved in a circular arc? Most people, after giving this careful consideration, will answer no, but they are wrong. There is a third curve that is self-congruent: the cir-cular helix. This is a curve that coils around a circular cylinder in such a way that it crosses the "elements" of the cylinder at a constant angle. Figure 1 makes this clear. The elements are the vertical lines that parallel the cylinder's axis; A is the constant angle with which the helix crosses every element. Because of the constant curvature of the helix a helical sword would screw its way easily in and out of a helical scabbard.

Actually the straight line and the circle can be regarded as limiting cases of the circular helix. Compress the curve until the coils are very close together and you get a tightly wound helix resembling a Slinky toy; if angle A increases to 90 de-

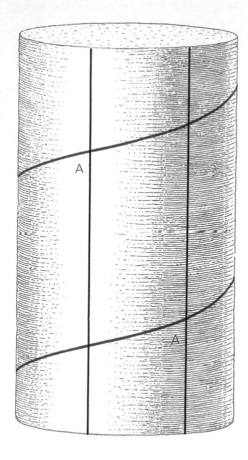

1. Circular helix (colored) *on cylinder*

grees, the helix collapses into a circle. On the other hand, if you stretch the helix until angle A becomes zero, the helix is transformed into a straight line. If parallel rays of light shine perpendicularly on a wall, a circular helix held before the wall with its axis parallel to the rays will cast on the wall a shadow that is a single circle. If the helix is held at right angles to the rays, the shadow is a sine curve. Other kinds of projections produce the cycloid and other familiar curves.

Every helix, circular or otherwise, is an asymmetric space curve that differs from its mirror image. We shall use the term "right-handed" for the helix that coils clockwise as it "goes away," in the manner of an ordinary wood screw or a corkscrew. Hold such a corkscrew up to a mirror and you will see that its reflection, in the words of Lewis Carroll's Alice, "goes the other way." The reflection is a left-handed corkscrew. Such a corkscrew actually can be bought as a practical joke. So unaccustomed are we to left-handed screw threads that a victim may struggle for several minutes with such a corkscrew before he realizes that he has to turn it counterclockwise to make it work.

Aside from screws, bolts, and nuts, which are (except for special purposes) standardized as right-handed helices, most man-made helical structures come in both right and left forms: candy canes, circular staircases, rope and cable made of twisted strands, and so on. The same variations in handedness are found in conical helices (curves that spiral around cones), including bedsprings and spiral ramps such as the inverted conical ramp in Frank Lloyd Wright's Guggenheim Museum in New York City.

Not so in nature! Helical structures abound in living forms, from the simplest virus to parts of the human body, and in almost every case the genetic code carries information that tells each helix precisely "which way to go." The genetic code itself, as everyone now knows, is carried by

2. *Helical horns of the Pamir sheep have opposite handedness*

a double-stranded helical molecule of DNA, its two right-handed helices twining around each other like the two snakes on the staff of Hermes. Moreover, since Linus Pauling's pioneer work on the helical structure of protein molecules, there has been increasing evidence that every giant protein molecule found in nature has a "backbone" that coils in a right-handed helix. In both nucleic acid and protein, the molecule's backbone is a chain made up of units each one of which is an asymmetric structure of the same handedness. Each unit, so to speak, gives an additional twist to the chain, in the same direction, like the steps of a helical staircase.

Larger helical structures in animals that have bilateral symmetry usually come in mirror-image pairs, one on each side of the body. The horns of rams, goats, antelopes, and other mammals are spectacular examples [*see Figure 2*]. The cochlea of the human ear is a conical helix that is left-handed in the left ear and right-handed in

the right. A curious exception is the tooth of the narwhal, a small whale that flourishes in arctic waters. This whimsical creature is born with two teeth in its upper jaw. Both teeth remain permanently buried in the jaw of the female narwhal, and so does the right tooth of the male. But the male's left tooth grows straight forward, like a javelin, to the ridiculous length of eight or nine feet—more than half the animal's length from snout to tail! Around this giant tooth are helical grooves that spiral forward in a counterclockwise direction [*see Figure 3*]. On the rare occasions when both teeth grow into tusks, one would expect the right tooth to spiral clockwise. But no, it too is always left-handed. Zoologists disagree on how this could come about. Sir D'Arcy Thompson, in his book *On Growth and Form*, defends his own theory that the whale swims with a slight screw motion to the right. The inertia of its huge tusk would produce a torque at the base of the tooth that might cause it to rotate counter-

3. *Helical grooves of the narwhal tooth are always left-handed*

clockwise as it grows (see "The Horn of the Unicorn," by John Tyler Bonner; *Scientific American*, March, 1951).

Whenever a single helix is prominent in the structure of any living plant or animal, the species usually confines itself to a helix of a specific handedness. This is true of countless forms of helical bacteria as well as of the spermatozoa of all higher animals. The human umbilical cord is a triple helix of one vein and two arteries that invariably coil to the left. The most striking instances are provided by the conical helices of the shells of snails and other mollusks. Not all spiral shells have a handedness. The chambered nautilus, for instance, coils on one plane; like a spiral nebula, it can be sliced into identical left and right halves. But there are thousands of beautiful molluscan shells that are either left- or right-handed [*see Figure 4*]. Some species are always left-handed and some always right-handed. Some go one way in one locality and the other way in another. Occasional "sports" that twist the wrong way are prized by shell collectors.

A puzzling type of helical fossil known as the devil's corkscrew (*Daemonelix*) is found in Nebraska and Wyoming. These huge spirals, six feet or more in length, are sometimes right-handed and sometimes left-handed. Geologists argued for decades over whether they are fossils of extinct plants or helical burrows made by ancestors of the beaver. The beaver theory finally prevailed after remains of small prehistoric beavers were found inside some of the corkscrews.

In the plant world helices are common in the structure of stalks, stems, tendrils, seeds, flowers, cones, leaves—even in the spiral arrangement of leaves and branches around a stalk. The number of turns made along a helical path, as you move from one leaf to the leaf directly above it, tends to be a number in the familiar Fibonacci series: 1, 2, 3, 5, 8, 13 . . . (Each number is the sum of the preceding two numbers.) A large literature in the field known as "phyllotaxy" (leaf arrangement) deals with the surprising appearance of the Fibonacci numbers in botanical phenomena of this sort.

The helical stalks of climbing plants are usually right-handed, but thousands of species of twining plants go the other way.

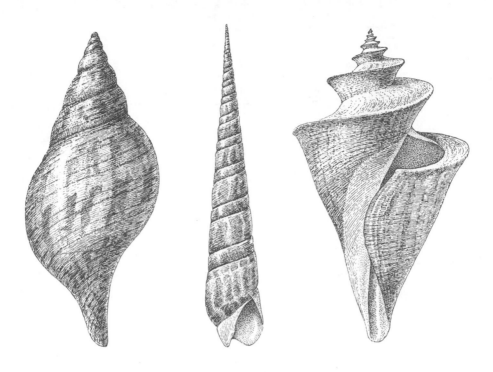

4. *Three molluscan shells that are right-handed conical helices*

The honeysuckle, for instance, is always left-handed; the bindweed (a family that includes the morning glory) is always right-handed. When the two plants tangle with each other, the result is a passionate, violent embrace that has long fascinated English poets. "The blue bindweed," wrote Ben Jonson in 1617, "doth itself enfold with honeysuckle." And Shakespeare, in *A Midsummer Night's Dream*, has Queen Titania speak of her intention to embrace Bottom the Weaver (who has been transformed into a donkey) by saying: "Sleep thou, and I will wind thee in my arms./ . . . So doth the woodbine the sweet honeysuckle/ Gently entwist." In Shakespeare's day "woodbine" was a common term for bindweed. Because it later came to be applied exclusively to honeysuckle many commentators reduced the passage to absurdity by supposing that Titania was speaking of honeysuckle twined with honeysuckle. Awareness of the opposite handedness of bindweed and honeysuckle heightens, of course, the meaning of Titania's metaphor.

More recently, a charming song called "Misalliance," celebrating the love of the honeysuckle for the bindweed, has been

MISALLIANCE

The fragrant Honeysuckle spirals clockwise to the sun
And many other creepers do the same.
But some climb counterclockwise, the Bindweed does, for one,
Or *Convolvulus,* to give her proper name.

Rooted on either side a door, one of each species grew,
And raced toward the window ledge above.
Each corkscrewed to the lintel in the only way it knew,
Where they stopped, touched tendrils, smiled and fell in love.

Said the right-handed Honeysuckle
To the left-handed Bindweed:
"Oh, let us get married,
If our parents don't mind. We'd
Be loving and inseparable.
Inextricably entwined, we'd
Live happily ever after,"
Said the Honeysuckle to the Bindweed.

To the Honeysuckle's parents it came as a shock.
"The Bindweeds," they cried, "are inferior stock.
They're uncultivated, of breeding bereft.
We twine to the right and they twine to the left!"

Said the counterclockwise Bindweed
To the clockwise Honeysuckle:
"We'd better start saving—
Many a mickle maks a muckle—
Then run away for a honeymoon
And hope that our luck'll
Take a turn for the better,"
Said the Bindweed to the Honeysuckle.

A bee who was passing remarked to them then:
"I've said it before, and I'll say it again:
Consider your offshoots, if offshoots there be.
They'll never receive any blessing from me."

Poor little sucker, how will it learn
When it is climbing, which way to turn?
Right—left—what a disgrace!
Or it may go straight up and fall flat on its face!

Said the right-hand-thread Honeysuckle
To the left-hand-thread Bindweed:
"It seems that against us all fate has combined.
Oh my darling, oh my darling,
Oh my darling Columbine,
Thou art lost and gone forever,
We shall never intertwine."

Together they found them the very next day.
They had pulled up their roots and just shriveled away,
Deprived of that freedom for which we must fight,
To veer to the left or to veer to the right!

MICHAEL FLANDERS

written by the British poet and entertainer Michael Flanders and set to music by his friend Donald Swann. With Flanders' kind permission the entire song is reproduced on the opposite page. (Readers who would like to learn the tune can hear it sung by Flanders and Swann on the Angel recording of *At the Drop of a Hat,* their hilarious two-man revue that made such a hit in London and New York.) Note that Flanders' honeysuckle is right-handed, his bindweed left-handed. It is a matter of convention whether a given helix is called left- or right-handed. If you look at the point of a right-handed wood screw, you will see the helix moving toward you counterclockwise, so that it can just as legitimately be called left-handed. Flanders simply adopts the convention opposite to the one taken here.

The entwining of two circular helices of opposite handedness is also involved in a remarkable optical-illusion toy that was sold in this country in the 1930's. It is easily made by twisting together a portion of two wire coils of opposite handedness [*see Figure* 5]. The wires must be soldered to each other at several points to make a rigid structure. The illusion is produced by pinching the wire between thumb and forefinger of each hand at the left and right edges of the central overlap. When the hands are moved apart, the fingers and thumbs slide along the wire, causing it to rotate and create a barber's-pole illusion of opposite handedness on each side. This is continuously repeated. The wire seems to be coming miraculously out of the inexhaustible meshed portion. Since the neutrino and antineutrino are now known to travel with screw motions of opposite handedness, I like to think of this toy as demonstrating the endless production of neutrinos and their mirror-image particles.

The helical character of the neutrino's path results from the fusion of its forward motion (at the speed of light) with its "spin." Helical paths of a similar sort are traced by many inanimate objects and living things: a point on the propeller of a moving ship or plane, a squirrel running up or down a tree, Mexican free-tailed bats gyrating counterclockwise when they emerge from caves at Carlsbad, New Mexico. Conically helical paths are taken by whirlpools, water going down a drain, tornadoes, and thousands of other natural phenomena.

Writers have found helical motions useful on the metaphorical level. The progress of science is often likened to an inverted conical spiral: the circles growing larger and larger as science probes further into the unknown, always building upward on

5. *Helical toy that suggests the production of neutrinos*

the circles of the past. The same spiral, a dark, bottomless whirlpool into which an individual or humanity is sliding, has also been used as a symbol of pessimism and despair. This is the metaphor that closes Norman Mailer's book *Advertisements for Myself.* "Am I already on the way out?" he asks. Time for Mailer is a conical helix of water flushing down a cosmic drain, spinning him off "into the spiral of star-lit empty waters."

And now for a simple helix puzzle. A rotating barber's pole consists of a cylinder on which red, white, and blue helices are painted. The cylinder is four feet high. The red stripe cuts the cylinder's elements (vertical lines) with a constant angle of 60 degrees. How long is the red stripe?

The problem may seem to lack sufficient information for determining the stripe's length; actually it is absurdly easy when approached properly.

Answer

If a right triangle is wrapped around any type of cylinder, the base of the triangle going around the base of the cylinder, the triangle's hypotenuse will trace a helix on the cylinder. Think of the red stripe of the barber's pole as the hypotenuse of a right triangle, then "unwrap" the triangle from the cylinder. The triangle will have angles of 30 and 60 degrees. The hypotenuse of such a triangle must be twice the altitude. (This is easily seen if you place two such triangles together to form an equilateral triangle.) In this case the altitude is four feet, so that the hypotenuse (red stripe) is eight feet.

The interesting part of this problem is that the length of the stripe is independent not only of the diameter of the cylinder but also of the shape of its cross section. The cross section can be an irregular closed curve of any shape whatever; the answer to the problem remains the same.

References

[References at the ends of chapters are listed chronologically so that the latest can be easily identified and to facilitate the adding of new references.]

Design in Nature. Vol. II. J. Bell Pettigrew. London: Longmans, Green, and Co., 1908.

The Curves of Life. Theodore Andrea Cook. New York: Henry Holt, 1914.

Introduction to Geometry. H. S. M. Coxeter. New York: John Wiley and Sons, 1961.

The Double Helix. James D. Watson. New York: Atheneum, 1968.

2. Klein Bottles and Other Surfaces

Three jolly sailors from
 Blaydon-on-Tyne
They went to sea in a bottle by Klein.
Since the sea was entirely inside
 the hull
The scenery seen was exceedingly dull.

Frederick Winsor,
The Space Child's Mother Goose

TO A TOPOLOGIST a square sheet of paper is a model of a two-sided surface with a single edge. Crumple it into a ball and it is still two-sided and one-edged. Imagine that the sheet is made of rubber. You can stretch it into a triangle or circle, into any shape you please, but you cannot change its two-sidedness and one-edgedness. They are topological properties of the surface, properties that remain the same regardless of how you bend, twist, stretch, or compress the sheet.

Two other important topological invariants of a surface are its chromatic number and Betti number. The chromatic number is the maximum number of regions that can be drawn on the surface in such a way that each region has a border in common with every other region. If each region is given a different color, each color will border on every other color. The chromatic number of the square sheet is 4. In other words, it is impossible to place more than four differently colored regions on the square so that any pair has a boundary in common. The term "chromatic number" also designates the minimum number of colors sufficient to color any finite map on a given surface. It is not yet known if 4 is the chromatic number, in this map-coloring sense, for the square, tube, and sphere, but for all other surfaces considered in this chapter, it has been shown that the chromatic number is the same under both definitions.

The Betti number, named after Enrico Betti, a nineteenth-century Italian physicist, is the maximum number of cuts that can be made without dividing the surface into two separate pieces. If the surface has edges, each cut must be a "crosscut": one that goes from a point on an edge to another point on an edge. If the surface is closed (has no edges), each cut must be a "loop cut": a cut in the form of a simple closed curve. Clearly the Betti number of the square sheet is 0. A crosscut is certain to produce two disconnected pieces.

If we make a tube by joining one edge of the square to its opposite edge, we create a model of a surface topologically distinct from the square. The surface is still two-sided but now there are two separate edges, each a simple closed curve. The chromatic number remains 4 but the Betti number has changed to 1. A crosscut from one edge to the other, although it eliminates the tube, allows the paper to remain in one piece.

A third type of surface, topologically the same as the surface of a sphere or cube, is made by folding the square in half along a diagonal and then joining the edges. The surface continues to be two-sided but all edges have been eliminated. It is a closed surface. The chromatic number continues to be 4. The Betti number is back to 0: any loop cut obviously creates two pieces.

Things get more interesting when we join one edge of the square to its opposite edge but give the surface a half-twist before doing so. You might suppose that this cannot be done with a square piece of paper, but it is easily managed by folding the square twice along its diagonals, as shown in Figure 6. Tape together the pair of edges indicated by the arrow in the last drawing. The resulting surface is the familiar Mö-

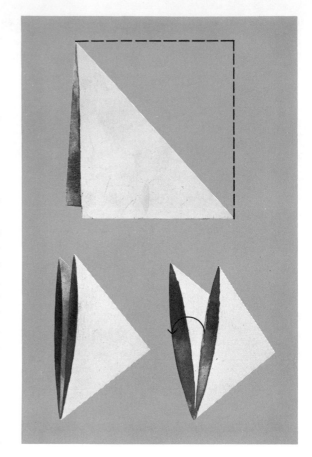

6. *Möbius surface constructed with a square*

bius strip, first analyzed by A. F. Möbius, the nineteenth-century German astronomer who was one of the pioneers of topology. The model will not open out, so it is hard to see that it is a Möbius strip, but careful inspection will convince you that it is. The surface is one-sided and one-edged, with a Betti number of 1. Surprisingly, the chromatic number has jumped to 6. Six regions, of six different colors, can be placed on the

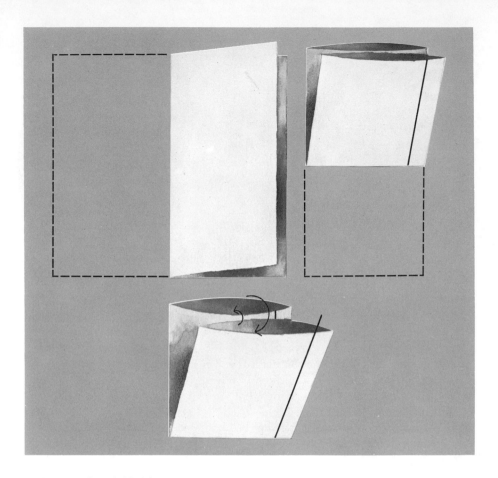

7. *Torus surface folded from a square*

surface so that each region has a border in common with each of the other five.

When both pairs of the square's opposite edges are joined, without twisting, the surface is called a torus. It is topologically equivalent to the surface of a doughnut or a cube with a hole bored through it. Figure 7 shows how a flat, square-shaped model of a torus is easily made by folding the square twice, taping the edges as shown by the solid gray line in the second drawing and the arrows in the last. The torus is two-sided, closed (no-edged) and has a chromatic number of 7 and a Betti number of 2. One way to make the two cuts is first to make a loop cut where you joined the last pair of edges (this reduces the torus to a tube) and then a crosscut where you joined the first pair. Both cuts, strictly speaking, are loop cuts when they are

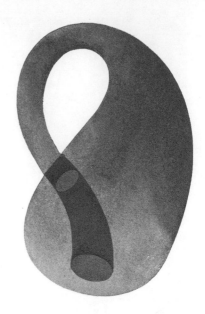

8. Klein bottle: a closed surface with no inside or outside

marked on the torus surface. It is only because you make one cut before the other that the second cut becomes a crosscut.

It is hard to anticipate what will happen when the torus model is cut in various ways. If the entire model is bisected by being cut in half either horizontally or vertically, along a center line parallel to a pair of edges, the torus surface receives two loop cuts. In both cases the resulting halves are tubes. If the model is bisected by being cut in half along either diagonal, each half proves to be a square. Can the reader find a way to give the model two loop cuts that will produce two separate bands interlocked like two rings of a chain?

Many different surfaces are closed like the surface of a sphere and a torus, yet one-sided like a Möbius strip. The easiest one to visualize is a surface known as the Klein bottle, discovered in 1882 by Felix Klein, the great German mathematician. An ordinary bottle has an outside and inside in the sense that if a fly were to walk from one side to the other, it would have to cross the edge that forms the mouth of the bottle. The Klein bottle has no edges, no inside or outside. What seems to be its inside is continuous with its outside, like the two apparent "sides" of a Möbius surface.

Unfortunately it is not possible to construct a Klein bottle in three-dimensional space without self-intersection of the surface. Figure 8 shows how the bottle is traditionally depicted. Imagine the lower end of a tube stretched out, bent up and plunged through the tube's side, then joined to the tube's upper mouth. In an actual model

made, say, of glass there would be a hole where the tube intersects the side. You must disregard this defect and think of the hole as being covered by a continuation of the bottle's surface. There is no hole, only an intersection of surfaces. This self-intersection is necessary because the model is in three-space. If we conceive of the surface as being embedded in four-space, the self-intersection can be eliminated entirely. The Klein bottle is one-sided, no-edged and has a Betti number of 2 and a chromatic number of 6.

Daniel Pedoe, a mathematician at Purdue University, is the author of *The Gentle Art of Mathematics*. It is a delightful book, but on page 84 Professor Pedoe slips into a careless bit of dogmatism. He describes

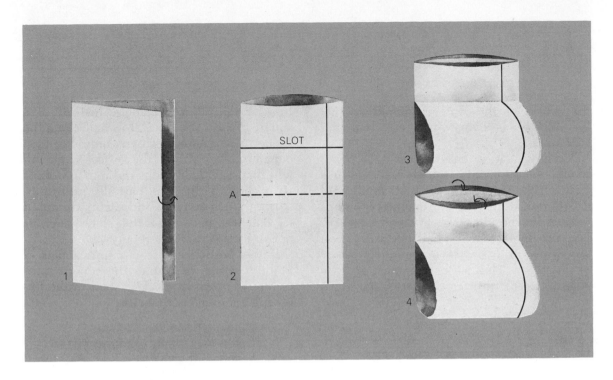

9. *Folding a Klein bottle from a square*

the Klein bottle as a surface that is a challenge to the glass blower, but one "which cannot be made with paper." Now, it is true that at the time he wrote this apparently no one had tried to make a paper Klein bottle, but that was before Stephen Barr, a science-fiction writer and an amateur mathematician of Woodstock, New York, turned his attention to the problem. Barr quickly discovered dozens of ways to make paper Klein bottles. Here I will describe only one of Barr's Klein bottles; one that enables us to continue working with a square and at the same time follows closely the traditional glass model.

The steps are given in Figure 9. First, make a tube by folding the square in half and joining the right edges with a strip of tape as shown [*Step 1*]. Cut a slot about a quarter of the distance from the top of the tube [*Step 2*], cutting only through the thickness of paper nearest you. This corresponds to the "hole" in the glass model. Fold the model in half along the broken line A. Push the lower end of the tube up through the slot [*Step 3*] and join the edges all the way around the top of the model [*Step 4*] as indicated by the arrows. It is not difficult to see that this flat, square model is topologically identical with the glass bottle shown in Figure 8. In one way it is superior: there is no actual hole. True, you have a slot where the surface self-intersects, but it is easy to imagine that the edges of the slot are joined so that the surface is everywhere edgeless and continuous.

Moreover, it is easy to cut this paper model and demonstrate many of the bottle's astonishing properties. Its Betti number of 2 is demonstrated by cutting the two loops formed by the two pairs of taped edges. If you cut the bottle in half vertically, you get two Möbius bands, one a mirror image of the other. This is best demonstrated by making a tall, thin model [*see Figure 10*]

10. *Bisected bottle makes two Möbius strips*

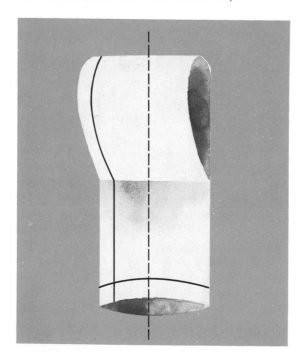

from a tall, thin rectangle instead of a square. When you slice it in half along the broken line (actually this is one long loop cut all the way around the surface), you will find that each half opens out into a Möbius strip. Both strips are partially self-intersecting, but you can slide each strip out of its half-slot and close the slot, which is not supposed to be there anyway.

If the bottle can be cut into a pair of Möbius strips, of course the reverse procedure is possible, as described in the following anonymous limerick:

> *A mathematician named Klein*
> *Thought the Möbius band was divine.*
> *Said he: "If you glue*
> *The edges of two,*
> *You'll get a weird bottle like mine."*

Surprisingly, it is possible to make a single loop cut on a Klein bottle and produce not two Möbius strips but only one. A great merit of Barr's paper models is that problems like this can be tackled empirically. Can the reader discover how the cut is made?

The Klein bottle is not the only simple surface that is one-sided and no-edged. A surface called the projective plane (because of its topological equivalence to a plane studied in projective geometry) is similar to the Klein bottle in both respects as well as in having a chromatic number of 6. As in the case of the Klein bottle, a model cannot be made in three-space without self-intersection. A simple Barr method for folding such a model from a square is shown in Figure 11. First cut the square along the

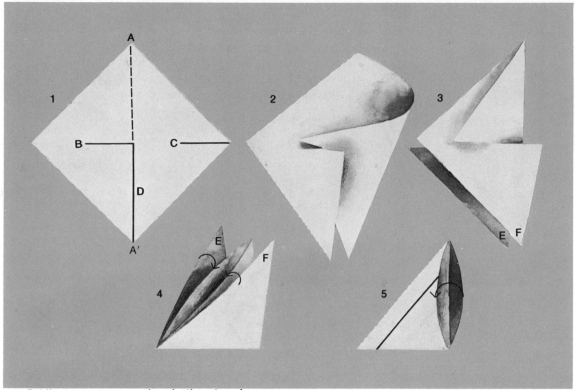

11. *Folding a cross-cap and projective plane from a square*

solid black lines shown in Step 1. Fold the square along the diagonal *A-A'*, inserting slot *C* into slot *B* [*Steps 2 and 3*]. You must think of the line where the slots interlock as an abstract line of self-intersection. Fold up the two bottom triangular flaps *E* and *F*, one on each side [*Step 4*], and tape the edges as indicated.

The model is now what topologists call a cross-cap, a self-intersecting Möbius strip with an edge that can be stretched into a circle without further self-intersection. This edge is provided by the edges of cut *D*, originally made along the square's diagonal. Note that unlike the usual model of a Mö-

bius strip, this one is symmetrical: neither right- nor left-handed. When the edge of the cross-cap is closed by taping it [*Step 5*], the model becomes a projective plane. You might expect it to have a Betti number of 2, like the Klein bottle, but it does not. It has a Betti number of 1. No matter how you loop-cut it, the cut produces either two pieces or a piece topologically equivalent to a square sheet that cannot be cut again without making two pieces. If you remove a disk from anywhere on the surface of the projective plane, the model reverts to a cross-cap.

Figure 12 summarizes all that has been

12. Topological invariants of seven basic surfaces

SURFACE		CHROMATIC NUMBER	SIDES	EDGES	BETTI NUMBER
SQUARE (OR DISK)		4	2	1	0
TUBE		4	2	2	1
SPHERE		4	2	0	0
MÖBIUS STRIP		6	1	1	1
TORUS		7	2	0	2
KLEIN BOTTLE		6	1	0	2
PROJECTIVE PLANE		6	1	0	1

said. The square diagrams in the first column show how the edges join in each model. Sides of the same color join each to each, with the direction of their arrows coinciding. Corners labeled with the same letter are corners that come together. Broken lines are sides that remain edges in the finished model. Next to the chromatic number of each model is shown one way in which the surface can be mapped to accommodate the maximum number of colors. It is instructive to color each sheet as shown, coloring the regions on both sides of the paper (as though the paper were cloth through which the colors soaked), because you must think of the sheet as having zero thickness. An inspection of the final model will show that each region does indeed border on every other one.

Answers

The torus-cutting problem is solved by first ruling three parallel lines on the unfolded square [*see Figure 13*]. When the square is folded into a torus, as explained, the lines make two closed loops. Cutting these loops produces two interlocked bands, each two-sided with two half-twists.

How does one find a loop cut on the Klein bottle that will change the surface to a single Möbius strip? On both left and right sides of the narrow rectangular model described you will note that the paper is creased along a fold that forms a figure-eight loop. Cutting only the left loop transforms the model into a Möbius band;

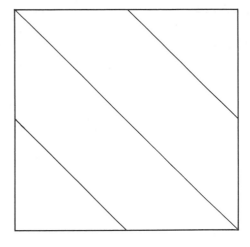

13. Solution to the torus-cutting problem

cutting only the right loop produces an identical band of opposite handedness.

What happens if both loops are cut? The result is a two-sided, two-edged band with four half-twists. Because of the slot the band is cut apart at one point, so that you must imagine the slot is not there. This self-intersecting band is mirror-symmetrical, neither right- nor left-handed. You can free the band of self-intersection by sliding it carefully out of the slot and taping the slot together. The handedness of the resulting band (that is, the direction of the helices formed by its edges) depends on whether you slide it out to the right or the left. This and the previous cutting problems are based on paper models that were invented by Stephen Barr and are described in his *Experiments in Topology.*

References

"Topology." Albert W. Tucker and Herbert S. Bailey, Jr. *Scientific American,* Vol. 182, No. 1; January, 1950. Pages 18–24.

Elementary Point Set Topology. R. H. Bing. *The American Mathematical Monthly,* Vol. 67, No. 7; August-September, 1960. Special Supplement.

Intuitive Concepts in Elementary Topology. Bradford Henry Arnold. New York: Prentice-Hall, 1962.

Experiments in Topology. Stephen Barr. New York: Thomas Y. Crowell, 1964.

Visual Topology. W. Lietzmann. London: Chatto and Windus, 1965.

The Four-Color Problem. Oystein Ore. New York: Academic Press, 1967.

3.   Combinatorial Theory

"AMID THE ACTION and reaction of so dense a swarm of humanity," Sherlock Holmes once remarked in reference to London, "every possible combination of events may be expected to take place, and many a little problem will be presented which may be striking and bizarre. . . ." Substitute "mathematical elements" for "humanity" and the great detective's remark is not a bad description of combinatorial mathematics.

In the language of set theory, combinatorial analysis is concerned with the arrangement of elements (discrete things) into sets, subject to specified conditions. A person playing chess is faced with a combinatorial problem: how best to bring about an arrangement of elements (chess pieces) on an eight-by-eight lattice, subject to chess rules, so that a certain element (his opponent's king) will be unable to avoid capture. A composer of music faces a combinatorial problem: how to arrange his elements (tones) in such a way as to arouse aesthetic pleasure. In the broadest sense, combinatorial tasks abound in daily life: seating guests around a table, solving crossword puzzles, playing card games, making out schedules, opening a safe, dialing a telephone number. When you put a key in a cylinder lock, you are using a mechanical device (the key) to solve the combinatorial problem of raising five little pins to the one permutation of heights that allows the cylinder to rotate. (This basic idea, by the way, goes back to wooden cylinder locks of ancient Egypt.)

Combinatorial number problems are as old as numbers. In China a thousand years before Christ mathematicians were exploring number combinations and permutations. The *Lo Shu*, an ancient Chinese magic square, is an exercise in elementary combinations. How can the nine digits be placed in a square array to form eight intersecting sets of three digits (rows, columns, and main diagonals), each summing to the same number? Not counting rotations and reflections, the *Lo Shu* [see Figure 18, page 24] is the only answer. It is a pleasant exercise in combinatorial thinking to see how simply

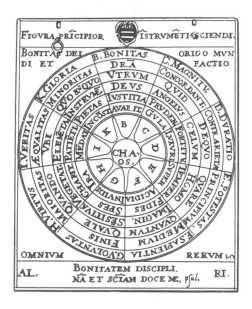

14. Two of Ramón Lull's combinatorial wheels

you can prove the *Lo Shu* pattern to be unique. (A good proof is given by Maurice Kraitchik in his *Mathematical Recreations*; New York: Dover, 1953; pages 146–147.)

In the thirteenth century Ramón Lull, an eccentric Spanish theologian, built a flourishing cult around combinatorial thinking. It was Lull's fervid conviction that every branch of knowledge could be reduced to a few basic principles and that by exploring all possible combinations of these principles one could discover new truths. To aid the mind in such endeavors Lull used concentric disks mounted on a central pin. Around the rim of each disk he placed letters symbolizing the basic ideas of the field under investigation; by turning the wheels one could run through all combinations of ideas. [*see Figure 14*]. Even today

there are survivals of Lullism in techniques developed for "creative thinking."

Until the nineteenth century most combinatorial problems were, like magic squares, studied as either mystical lore or mathematical recreations. To this day they provide a large share of puzzle problems, some of which are trivial brain teasers: A drawer contains two red socks, two green socks, and two blue socks. What is the smallest number of socks you can take from the drawer, with your eyes closed, and be sure you have a pair that matches?

There are moderately difficult questions such as: In how many different ways can a dollar be changed with an unlimited supply of halves, quarters, dimes, nickels, and pennies?

And there are problems so difficult they

have not yet been solved: Find a formula for the number of different ways a strip of *n* postage stamps can be folded. Think of the stamps as being blank on both sides. Two ways are not "different" if one folded packet can be turned in space so that its structure is the same as the other. Two stamps can be folded in only one way, three stamps in two ways, four in five ways [*see Figure 15*]. Can the reader give the number of different ways a strip of five stamps can be folded?

It was not until about 1900 that combinatorial analysis began to be recognized as an independent branch of mathematics, and not until the 1950's that it suddenly grew into a vigorous new discipline. There are many reasons for this upsurge of interest. Modern mathematics is much concerned with logical foundations, and a large part of formal logic is combinatorial. Modern science is much concerned with probability, and most probability problems demand prior combinatorial analysis. Almost everywhere science looks today it discovers not continuity but discreteness: molecules, atoms, particles, the quantum numbers for charge, spin, parity, and so on. Wolfgang Pauli's "exclusion principle," which finally explained the structure of the periodic table of elements, was the outcome of combinatorial thinking.

The great revolution that is now under way in biology springs from the sensational discovery that genetic information is carried by a nucleic acid code of four letters taken three at a time in a way that recreational mathematicians have been exploring for

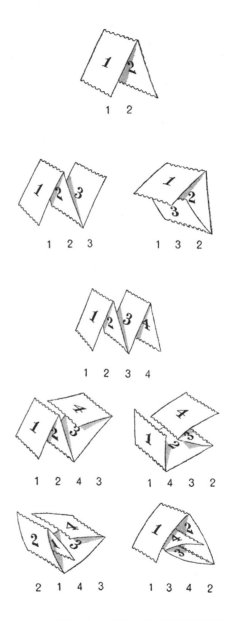

15. Ways of folding two, three, and four stamps

more than a century. Perhaps it is no accident that the first suggestion the genetic code consisted of triplets of four symbols was made by the physicist George Gamow, who always had a keen interest in mathematical puzzles. (For the story of this remarkable insight, see the afterword of Gamow's autobiography, *My World Line*.) Information theory with its bits and code words, computers with their yes and no circuits raise a myriad of combinatorial questions. At the same time the computer has made possible the solution of combinatorial problems that had previously been too complex to solve. This too has surely been a factor in stimulating interest in combinatorial mathematics.

The two main types of combinatorial problem are "existence" problems and "enumeration" problems. An existence problem is simply the question of whether or not a certain pattern of elements exists. It is answered with an example or a proof of possibility or impossibility. If the pattern exists, enumeration problems follow. How many varieties of the pattern are there? What is the best way to classify them? What patterns meet various maxima and minima conditions? And so on.

We can illustrate both types of problem by considering the following simple question: Is it possible to arrange a set of positive integers from 1 to n in a hexagonal array of n cells so that all rows have a constant sum? In short: Is a magic hexagon possible?

The simplest such array of cells is shown in Figure 16. Can the digits from 1 to 7 be

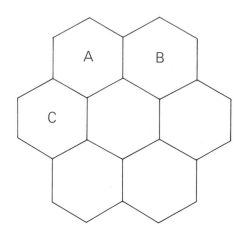

16. "Order 2" magic hexagon impossibility proof

placed in those seven cells in such a way that each of the nine rows has the same sum? The sum, called the magic constant, is easily determined. We have only to add the digits from 1 to 7 and then divide by 3 — the number of rows that are parallel in a given direction. The sum is 28, but it is not evenly divisible by 3. Since the magic constant must be an integer, we have proved that an "order 2" magic hexagon (the order is the number of cells on a side) is impossible. For an even simpler impossibility proof consider corner-cell A. It belongs to two rows that contain only two cells. If both rows have the same sum, cells B and C will have to contain the same digit, but this violates a condition of the problem that was given.

Turning attention to the next largest array, an order-3 hexagon with 19 cells, we find that the numbers sum to 190 — which *is* divisible by 5, the number of parallel rows in one direction. The magic constant is 38.

The previous impossibility proof has failed, but of course this does not guarantee that an order-3 magic hexagon exists.

In 1910 Clifford W. Adams, now living in Philadelphia as a retired clerk for the Reading Railroad, began searching for a magic hexagon of order 3. He had a set of hexagonal ceramic tiles made, bearing the numbers 1 to 19, so that he could push them around and explore patterns easily. For forty-seven years he worked at the task in odd moments. In 1957, convalescing from an operation, he found a solution [*see Figure 17*]. He jotted it down on a sheet of paper but mislaid the sheet, and for the next five years he tried in vain to reconstruct

17. The only possible magic hexagon

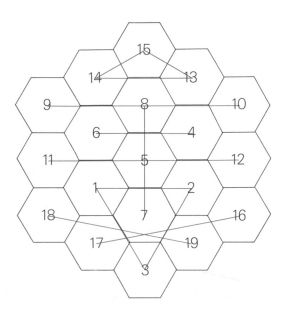

his solution. In December 1962 he found the paper, and early the following year he sent me the pattern. Each of the 15 rows sums to 38. The colored lines connect consecutively numbered cells in sets of twos and threes to bring out the pattern's curious bilateral symmetry. (A similar symmetry is displayed by the *Lo Shu* if it is tipped so cell 2 is at the top, cell 8 is at the bottom, and triplets (1, 2, 3), (4, 5, 6), and (7, 8, 9) are joined by lines.)

When I received this hexagon from Adams, I was only mildly impressed. I assumed that there was probably an extensive literature on magic hexagons and that Adams had simply discovered one of hundreds of order-3 patterns. To my surprise a search of the literature disclosed not a single magic hexagon. I knew that there were 880 different varieties of magic squares of order 4, and that order-5 magic squares have not yet been enumerated because their number runs into the millions. It seemed strange that nothing on magic hexagons had been published.

I sent the Adams hexagon to Charles W. Trigg, a mathematician at Los Angeles City College who is an expert on combinatorial problems of this sort. A post-card reply confirmed the hexagon's unfamiliarity. A month later I was staggered to receive from Trigg a formal proof that no other magic hexagon of *any* size is possible. Among the infinite number of ways to place integers from 1 to *n* in hexagonal arrays, only *one* pattern is magic!

Trigg's proof of impossibility for orders above 3 calls on Diophantine analysis, the obtaining of integral solutions for equations. Trigg first worked out the formula for the magic constant in terms of order n:

$$\frac{9(n^4 - 2n^3 + 2n^2 - n) + 2}{2(2n - 1)}$$

This is easily changed to an equation in which $5/(2n-1)$ is an integral term. To be integral, n must be either 1 or 3. A magic hexagon of one cell is of course trivial. Adams had found one pattern for order 3. Are there other arrangements of the 19 integers (not counting rotations and reflections) that are magic? Trigg's negative answer was obtained by combining brute force (he used a ream and a half of sheets on which the cell pattern had been reproduced six times) with clever short cuts. His result was later verified by numerous computer programs. (Trigg explained his proof, and discussed curious properties of the hexagon, in "A Unique Magic Hexagon," *Recreational Mathematics Magazine*, January, 1964.)

As an elementary exercise the reader is invited to see if he can rearrange the 19 digits in Adams' hexagon so that the pattern is magic in the following way: each 3-cell row adds to 22, each 4-cell row to 42, each 5-cell row to 62. Magic hexagons of *this* type have been explored before and there are large numbers of them. (The problem is solved easily with the right insight. *Hint:* The new pattern can be obtained by applying the same simple transformation to each number.)

A pattern of integers arranged in a unique, elegant manner usually has many bizarre properties. Even the ancient *Lo Shu* still harbors surprises. A few years ago Leo Moser of the University of Alberta discovered an amusing paradox that arises when the *Lo Shu* is regarded as a chart of the relative strengths of nine chess players [*see Figure 18*]. Let row A be a team of three chess experts with the playing strengths of 4, 9, and 2 respectively. Rows B and C are two other teams, with playing strengths as indicated. If teams A and B play a round-robin tournament, in which every player of one team plays once against every player of the other team, team B will win five games and team A will win four. Clearly team B is stronger than A. When team B plays team C, C wins five games and loses four, so that C is obviously stronger than B. What happens when C, the

18. The Lo Shu, *ancient Chinese magic square*

A	4	9	2
B	3	5	7
C	8	1	6

strongest team, plays A, the weakest? Work it out yourself. Team A is the winner by five to four! Which, then, is the strongest team? The paradox brings out the weakness of round-robin play in deciding the relative strengths of teams. Moser has analyzed many paradoxes of this sort, of which this is one of the simplest. The paradox also holds if teams A, B, and C are the columns of the *Lo Shu* instead of the rows.

Similar paradoxes, Moser points out, arise in voting. For example, assume that one person's preference for three candidates is in the order A, B, C. A second person prefers B, C, A and a third prefers C, A, B. It is easy to see that a majority of the three voters prefers A to B, a majority prefers B to C, and (confusingly) a majority also prefers C to A! This simple paradox was apparently first discussed in 1785 by the French mathematician, the Marquis de Cordorcet, and first rediscovered by Lewis Carroll who published several remarkable pamphlets on voting procedures. The paradox was independently rediscovered later by many others. (For a history of the paradox, and a listing of important recent works in which its implications for group decision theory are analyzed, see "Voting and the Summation of Preferences," by William H. Riker, *The American Political Science Review,* December, 1961. On the application of the paradox to the scores of competing teams, see "A Paradox in the Scoring of Competing Teams," by E. V. Huntington, *Science,* Vol. 88, 1938, pages 287–288.)

The arrangement of elements in square and rectangular matrices provides a large portion of modern combinatorial problems, many of which have found useful applications in the field of experimental design. In Latin squares the elements are so arranged that an element of one type appears no more than once in each row and column. Here is a pretty combinatorial problem along such lines that is not difficult but conceals a tricky twist that may escape many readers:

Suppose you have on hand an unlimited supply of postage stamps with values of one, two, three, four, and five cents (that is, an unlimited supply of each value). You wish to arrange as many stamps as possible on a four-by-four square matrix so that no two stamps of the same value will be in the same row, column, or any diagonal (not just the two main diagonals). In other words, if you place a chess queen on any stamp in the square and make a single move in any direction, the queen's path will not touch two stamps of like value. There is one further proviso: the total value of the stamps in the square must be as large as possible. What is the maximum? No cell may contain more than one stamp, but one or more cells may, if you wish, remain empty.

Addendum

After my publication of the magic hexagon, John R. A. Cooper called my attention to a prior publication without commentary by Tom Vickers in *The Mathematical Gazette,* December, 1958, page 291. So far as I know, this was the first appearance of the hexagon in print. Karl Fabel sent me a letter he had received from Martin Kühl, of Hanover, Germany, showing that Kühl, too, had independently discovered the hexagon (about 1940) but had not published it.

A feature story by Karl Abraham on Clifford W. Adams' discovery of the pattern appeared in the Philadelphia *Evening Bulletin*, July 19, 1963, page 18; a follow-up story giving the solution (which readers had been asked to find) appeared in the July 30 issue.

Answers

The combinatorial questions are answered as follows:

Four socks guarantee a matching pair.

A dollar can be changed in 292 distinct ways. For a full solution, using recursive computation, see the last two pages of George Polya's *How to Solve It*; Second edition; New York: Doubleday, 1957.

A strip of five stamps, blank on both sides, can be folded in 14 distinct ways [*see Figure 19*]. (If the stamps are printed on one side, you might think the number of ways would double, but it increases only to 25. Why?)

The problem of finding a formula for *n* stamps remains unsolved, but recursive procedures by which the number of differ-

19. Answer to the stamp-folding problem

ent folds can be calculated easily with computers have been developed. This problem was first posed by S. M. Ulam. In 1961 Mark B. Wells, using a computer at Los Alamos Scientific Laboratory, found the number of distinct foldings for six, seven, eight, and nine stamps to be 38, 120, 353, and 1,148 respectively. More recent results will be found in "A Map-folding Problem," by W. F. Lunnon, in *Mathematics of Computation,* January, 1968; and in "Folding a Strip of Stamps," by John E. Koehler, in *Journal of Combinatorial Theory,* September, 1968. The problem of finding a nonrecursive formula is more difficult, and also unsolved, if one asks for different ways of folding square sheets into a packet of unit squares.

To change the magic hexagon to a hexagon with 22 as the sum of each three-cell row, 42 as the sum of each four-cell row, and 62 as the sum of each five-cell row, replace the number in each cell with the difference between that number and 20.

The problem of placing stamps with values of one, two, three, four, and five cents in a four-by-four square, with no two stamps of the same value in any row, column, or diagonal (including the smaller diagonals), can be answered with a maximum value of 50 cents [*see Figure 20, which shows one of many solutions*]. This is probably two cents more than most readers were able to achieve if they used four fours and left two cells empty. The trick is to use only three four-cent stamps. "The reader will probably find, when he sees the solution," wrote Henry Dudeney in

20. *A solution to the stamp-placing problem*

Amusements in Mathematics (Problem 308), "that, like the stamps themselves, he is licked."

Donald E. Knuth found that Dudeney's solution could be enlarged to a remarkable five-by-five square, satisfying all the conditions and giving the maximum total value of 75 cents. Simply add a top row of 2, 1, 4, 3, 5, and a right border (reading downward) of 5, 1, 3, 2, 4. Each value appears five times in the square. This is equivalent to the problem of superimposing five solutions to the problem of the non-attacking queens on the order-5 board. (See Chapter 16 of my *Unexpected Hanging;* New York: Simon and Schuster, 1969.)

References

Combinatorial Analysis. Percy A. MacMahon. Cambridge: Cambridge University Press, Vol. 1, 1915; Vol. 2, 1916. (Reprint [2 vols. in one]. Bronx, N.Y.: Chelsea Publishing Co., 1960.)

An Introduction to Combinatorial Analysis. John Riordan. New York: John Wiley and Sons, 1958.

Combinatorial Mathematics. Herbert John Ryser. New York: John Wiley and Sons, 1963.

Applied Combinatorial Mathematics. Frank Harary. New York: John Wiley and Sons, 1964.

Combinatorial Theory. Marshall Hall, Jr. Waltham, Mass.: Blaisdell Publishing Co., 1967.

Elementary Combinatorial Analysis. Martin Eisen. New York: Gordon and Breach, 1969.

4. Bouncing Balls in Polygons and Polyhedrons

THROUGHOUT recorded history the bouncing ball has been indispensable equipment for a dazzling variety of indoor and outdoor sports. Games exploiting it range from the child's simple bouncing of a rubber ball ("One, two, three O'Lary . . .") to sports such as tennis, handball, and billiards in which the ability to judge angles of incidence and reflection is essential to a player's skill.

> *The balls*
> *shine round and clear, quick blobs*
> *of color on faultless fields,*
> *where rapid vengeance rolls*
> *and clicks, returns*
> *or poorly judged, deflects*
> *to pass and spend itself in motion*
> *rebounding gingerly from cushions . . .*
>
> Herman Spector,
> "B.A. (Billiard Academy)"

Mathematicians and physicists are notoriously fond of pool and billiards. It is easy to understand why. The gingerly rebounds within faultless fields can be precisely calculated. Lewis Carroll, who taught mathematics at the University of Oxford, enjoyed playing billiards, particularly on a *circular* table he had made for himself. A much prized collector's item is the first edition of a two-page leaflet, published by Carroll in 1890 and never reprinted, that explains his rules for this game.

Hundreds of recreational problems concern the rebounds of elastic balls within perimeters of various shapes. Consider, for example, the following old puzzle: You have two vessels with respective capacities of 7 and 11 pints. Beside you is a large tub of water. Using only the two vessels (and excluding all dodges such as marking the containers or tilting them to obtain fractional amounts), how can you measure exactly two pints?

The question can be answered by trial and error or by applying various algebraic procedures. What has all this to do with bouncing balls? Surprisingly, liquid-measuring puzzles of this type can be solved easily by graphing the paths of balls bouncing inside *rhomboidal* tables! (The method, using what topologists call a "directed graph," was first explained by M. C. K. Tweedie in *The Mathematical Gazette* of July, 1939.) The cushions of such tables are best drawn on isometric graph paper: paper with a lattice of equilateral triangles. In this case the sides of the table are 7 and 11 units [*see Figure 21*]. Readings on the horizontal axis represent the amount of water in the 11-pint vessel at any time and

readings on the vertical axis tell how much water is in the 7-pint vessel.

To use the graph, imagine a ball at point *0* in the lower left corner. It travels to the right along the base of the rhomboid until it strikes the right-hand cushion at a point labeled *11* on the base line: the 11-pint vessel has been filled and the 7-pint container remains empty. After bouncing off the right-hand cushion the ball travels up and to the left until it hits the top cushion at point *4* on the horizontal co-ordinate

21. *Graph and 18-step solution of a liquid-pouring problem*

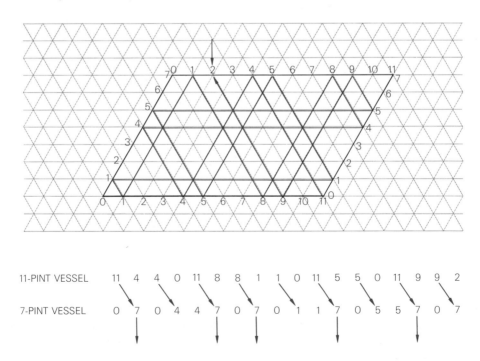

11-PINT VESSEL	11	4	4	0	11	8	8	1	1	0	11	5	5	0	11	9	9	2
7-PINT VESSEL	0	7	0	4	4	7	0	7	0	1	1	7	0	5	5	7	0	7

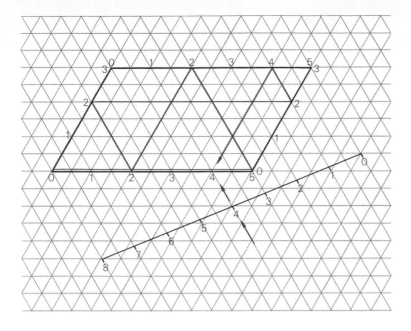

22. *Graph of Tartaglia's puzzle*

and on the seventh line on the side co-ordinate. This plot indicates that 7 pints have been transferred from the 11-vessel to the 7-vessel, leaving 4 pints in the larger vessel.

If you continue to follow the bouncing ball until it strikes a point marked 2, keeping a record of each step, you will obtain the 18-step answer shown below the graph. Slanting arrows indicate that water is poured from one vessel into another. The vertical arrows show either that the 7-vessel is being emptied into the tub or that the 11-vessel is being filled.

Is this the shortest answer? No; an alternative procedure is to begin by filling the 7-vessel. This is graphed by starting the ball at the 0 point and rolling it *up* along the table's left side. If the reader traces the ball's path until it strikes a 2 point, keeping a record of the steps, he will find that his

ball computer bounces out a solution in 14 steps—the minimum.

With a little ingenuity one can devise ball-bounce computers for any liquid-pouring puzzle in which no more than three vessels are involved. Consider the oldest of all three-vessel problems, which goes back to Nicola Fontana, the sixteenth-century Italian mathematician who called himself Tartaglia ("The Stammerer"). An eight-pint vessel is filled with water. By means of two empty vessels that hold five and three pints respectively, divide the eight pints evenly between the two larger vessels. The graph for this problem is shown in Figure 22. Here the eight-pint vessel is represented by a line paralleling a main diagonal of the rhomboid. The ball begins as before in the 0 corner. It is easy to trace a path that computes the minimum solution, which requires seven operations.

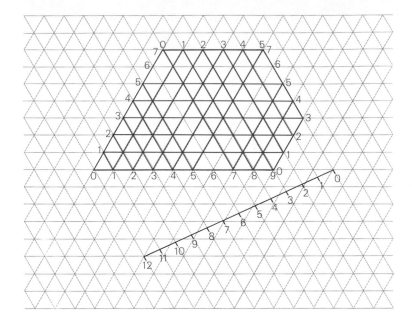

23. Graph for vessels
of volumes 7, 9, and 12

When the two smaller vessels have no common divisor and the third vessel is equal to or greater than the sum of the smaller vessels, it is possible to measure out any whole number from 1 to the capacity of the middle-sized vessel. For example, with vessels of 15-, 16-, and 31-pint capacities one can measure any quantity from 1 to 16. This is not possible if the two smaller vessels have a common divisor. A graph for vessels of 4, 6, 10 will not bounce the ball to any odd number, and vessels of 3, 9, 12 will measure only the quantities 3, 6, 9. (In both cases only multiples of the common divisor can be measured.) If the largest vessel is *smaller* than the sum of the other two, there are further limitations. For example, vessels of 7, 9, 12 require that a corner of the rhomboidal graph be sliced off [*see Figure 23*]. The bouncing ball will measure any quan-

tity from 1 to 9 except 6. Although 7 and 9 have no common divisor, the smallness of the third vessel makes it impossible to obtain 6.

When the largest vessel is *larger* than the sum of the other two, the graph continues to be applicable. The reader may enjoy applying it to the following variation of Tartaglia's problem, as posed by Sam Loyd on page 304 of his famous *Cyclopedia of Puzzles*. (This is one of the puzzles for which the *Cyclopedia* fails to furnish an answer, a fact that may explain why the puzzle has never been reprinted.)

Some U.S. soldiers managed to "capture" a 10-gallon keg of beer. "They naturally sampled a part of it," writes Loyd, making use of 3-gallon and 5-gallon containers. The rest of the beer was carried back to camp in three equal portions—one in the keg and the other two in the two containers. How much did they drink and how did they measure the remainder into three equal (non-zero) parts? The best solution is the one with the fewest steps for the entire procedure. Each step, including the drinking operation, involves an integral number of gallons, and it is assumed that no beer is wasted by being tossed out.

You may find it entertaining to experiment with vessels of various sizes, using the ball computer to explore all that can be done with them. For more information about the technique, including its extension to four vessels by means of tetrahedral graphs, the interested reader is referred to the book by T. H. O'Beirne listed at the close of this chapter.

A different type of ball-bouncing problem is that of finding cyclic paths along which a ball can bounce forever inside a polygon, always tracing the same path and hitting each side only once in each cycle. Such problems can be solved by using the powerful technique of mirror reflection. A table in the shape of a square provides a simple example. Figure 24 shows a square reflected along three different sides, and the colored line is its only cyclic path with segments of equal length. Folding the four squares into a unit-square packet transforms the straight line into the cyclic path.

At this point two interesting questions arise. Are there cyclic paths with equal segments inside the solid analogues of the square and equilateral triangle: the cube and tetrahedron? The ball is assumed to be an idealized elastic particle (or a light ray

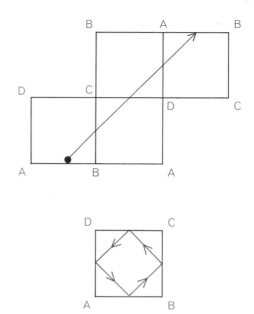

24. Equal-segment path in a square

inside a solid with interior mirror surfaces), taking straight paths in zero gravity and bouncing off the sides in the usual manner: with equal angles of incidence and reflection on a plane perpendicular to the side against which it bounces. The ball must strike each face only once during the cycle and travel the same distance between each consecutive pair of bounces. (Striking an edge or corner is not regarded as striking the faces meeting at that edge or corner; otherwise the cube problem would be solved by a ball moving back and forth between two diagonally opposite corners.)

Warren Weaver, in one of his many articles on Lewis Carroll, has disclosed that the cube problem is found among Carroll's unpublished and mathematical notes. It is the

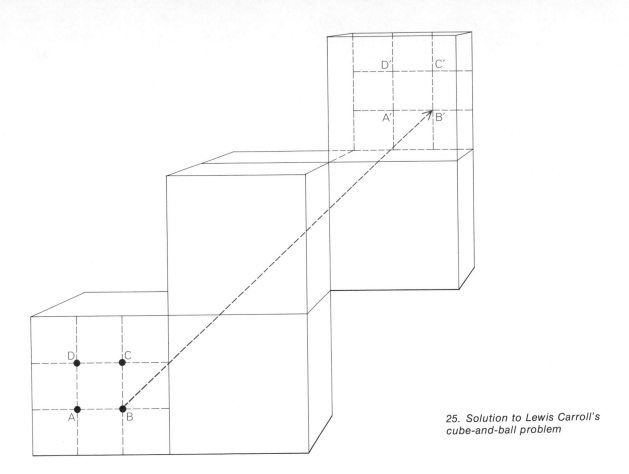

25. Solution to Lewis Carroll's
cube-and-ball problem

sort of problem that would appeal to the inventor of circular billiards. Actually the notion of playing billiards inside a cubical "table" is not as farfetched as it might seem. With gigantic space stations perhaps only a few decades away it takes no great prophetic ability to foresee a variety of three-dimensional sports that will take advantage of zero gravity. Pool adapts neatly to a rectangular room with cushion walls, floor and ceiling, corner pockets, and balls numbered from 1 to 35 that are initially arranged in tetrahedral formation. Of course, there

would be difficulties. Air resistance offers much less friction than the felt surface of a pool table does. If the tetrahedron were broken by a fast cue ball, entropy would increase at a rapid rate. It would be hard to keep out of the way of balls flying about in random directions like the molecules of a gas in thermal equilibrium!

But back to Carroll's problem. The reflection technique used with squares can be applied to cubes. Five reflections are required and the colored line in Figure 25 traces the desired path. It is one of four different paths, identical in shape, that solve the problem. (If all six faces of the cube are ruled into nine smaller squares, each path touches every face at one corner of the central square.) Figure 26 shows a cardboard model that demonstrates the path after the six cubes have been "folded" into one another. The cord is held in place by passing loops through small holes and securing them on the outside with pegs made of wood. If you think of the cube as being formed of 27 smaller cubes, you will see that every segment of the path is a diagonal of a small cube. Each segment therefore has a length of $1/\sqrt{3}$ on a unit cube. The path's total length is $2\sqrt{3}$.

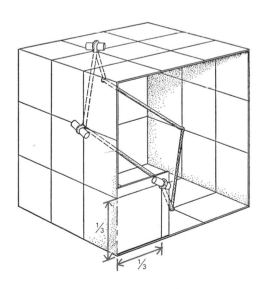

26. *Model showing path inside a cube*

As far as I know, Hugo Steinhaus was the first to find this path. (See his *One Hundred Problems in Elementary Mathematics;* New York: Basic Books, 1964, Problem 33. The book is a translation of the 1958 Polish edition.) The solution was later rediscovered by Roger Hayward, who published it in *Recreational Mathematics Magazine,* June, 1962. The shape of the path, he writes, is known to organic chemists as a "chair-shaped hexagon." It occurs often in carbon compounds, such as cyclohexane, in which six carbon atoms are single-bonded in a ring with other atoms attached outside the ring. "It is interesting to note," writes B. M. Oliver of the Hewlett-Packard Company in Palo Alto, California, "that the path appears as a 1×2 rectangle in all projections of the cube taken perpendicular to a face, as a rhombus in three of the isometric projections taken parallel to a diagonal of the cube, and as a regular hexagon in the fourth isometric view. A queer figure, but that's the way the ball bounces!"

A similar cyclic path inside a tetrahedron was discovered by John H. Conway and later, independently, by Hayward in 1962. It is easy to reflect a tetrahedron three times [*see Figure 27*] and find a cyclic path that touches each side once. The difficult trick is to find a cyclic path with equal segments. One is shown by the colored line. There are three such paths, all alike, touching each face of the solid at one corner of a small equilateral triangle in the center of the face. The side of this small triangle is a tenth of the edge of a tetrahedron with an edge of 1. Each segment of the ball's path has a length

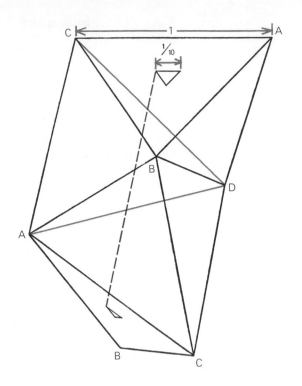

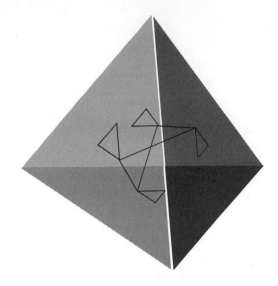

28. *Acetate model of a path in the tetrahedron*

27. *Solution to the problem of a ball
in the tetrahedron*

of $\sqrt{10}/10$, or .31622777+, giving the path a total length of 1.2649+.

Hayward made a handsome acetate model in which nylon thread traces the path of the bouncing ball (or light ray) after the four tetrahedrons have been "folded" together [*see Figure 28*]. He cut the sides from sheets of acetate and cemented them along their edges after drilling four small holes at the proper points. Before cementing the last side he looped the thread through the holes of three faces and held it with pieces of tape on the outside. The two free ends were drawn through the hole in the fourth face, which was then cemented to the other

three. After tightening the thread by pulling on the loops he sealed each hole with a drop of acetone mixed with Duco household cement and trimmed the outside loops and ends. A similar acetate model can be made of the cube. In both models threads of different colors can be used to show all possible paths.

Addendum

G. de Josselin de Jong of Holland generalized Lewis Carroll's ball-in-cube problem to hypercubes of all higher dimensions. He wrote out his analysis for me in 1963 and I do not know if he has since published it. The length of the ball's path is given by the simple formula $2\sqrt{n}$, where n is the number of dimensions. Therefore, the ball's path inside a unit four-space hypercube has a length of exactly four units. In all higher spaces the path is unique except, of course, for rotations and reflections.

Answer

Given a ten-gallon keg filled with beer and two vessels of three-gallon and five-gallon capacity, how can one (in the minimum number of operations) drink a quantity of beer and leave equal (nonzero) amounts in each of the three vessels? Since the vessels measure only integral amounts, the beer to be divided into thirds must be a multiple of

three: three, six or nine gallons. The first two amounts can be eliminated because in both cases a third of the amount is less than the capacity of each vessel. (After any pouring operation at least one vessel must be either empty or full. Neither situation would obtain if each vessel contained less than its capacity.) We conclude, therefore, that one gallon must be drunk, leaving nine to be divided into thirds.

The ball-bouncing computer traces a minimum path that measures one gallon [*see Figure 29*]. After the gallon (in the three-gallon vessel) is drunk, four gallons remain

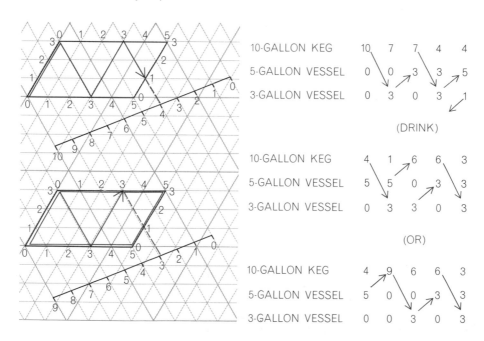

29. *Solution to Sam Loyd's problem*

in the ten-gallon keg, five in the five-gallon vessel. The three-gallon vessel is empty. This new situation is diagramed as shown in the lower graph. The ball must now reach a point that marks three gallons in each container. The minimum path is shown in color, with two alternative steps in a lighter shade. Counting the drinking of the gallon as an "operation," the complete solution involves nine operations, which are shown below the two graphs.

References

"A Graphical Method of Solving Tartaglian Measuring Puzzles." M. C. K. Tweedie. *The Mathematical Gazette,* Vol. 23, No. 255; July, 1939. Pages 278–282.

"A Billiard Ball Computer." Aaron Bakst. In his *Mathematical Puzzles and Pastimes.* New York: Van Nostrand Co., 1954. Chapter 2, Pages 10–21.

"Pouring Problems: The Robot Method." Nathan Altshiller Court. In his *Mathematics in Fun and in Earnest.* New York: Dial Press, 1958. Pages 223–231.

Graphs and Their Uses. Oystein Ore. New York: Random House, 1963. Pages 68–70.

"Billiard Balls in an Equilateral Triangle." Donald E. Knuth. *Recreational Mathematics Magazine,* No. 14; January, 1964. Pages 20–23.

Puzzles and Paradoxes. T. H. O'Beirne. New York: Oxford University Press, 1965. Chapter 4.

Mathematics on Vacation. Joseph S. Madachy. New York: Charles Scribner's Sons, 1966. Pages 231–241.

Geometry Revisited. H. S. M. Coxeter. New York: Random House, 1967. Pages 89–93.

Algorithms, Graphs, and Computers. Richard Bellman, Kenneth L. Cooke, and Jo Ann Lockett. New York: Academic Press, 1970. Chapter 5.

5. Four Unusual Board Games

DURING the 1960's there was a remarkable upsurge of interest in mathematical board games. Today more people than ever before are playing the traditional games such as chess and experimenting with the new games that keep turning up in the stores. More mathematicians are analyzing the strategies of such games and more computers are being programed to play them. In this chapter we examine four excellent but little-known board games, two new and two old. Their playing fields can easily be drawn on paper or cardboard, the rules of play are quite simple and everyone in the family will find the contests great fun.

The Military Game, as it is called in France, is a splendid example of a two-player game that combines extreme simplicity with extraordinary strategic subtlety. According to Édouard Lucas, who describes the game in Volume III (pages 105–116) of his celebrated *Récréations Mathématiques*, the game was popular in French military circles during and after the Franco-Prussian War of 1870–1871. It is a

pity that it has since been so completely forgotten; not one of the standard histories of board games even mentions it.

The board for the Military Game is shown in Figure 30 with the positions labeled to facilitate description. One player—we will call him White—has three men that are initially placed on the colored spots. *A*, *1*, and *3*. Black, his opponent, has only one man, which he places on spot *5* in the center. (Chess pawns can be used for men, or three pennies and a nickel.) White moves first and the game proceeds with alternate turns. Black may move in any direction along a line from one spot to a neighboring spot. White moves similarly, but only left, right or forward (straight ahead or diagonally), never backward. There are no captures. White wins if he can pin Black's piece so that it cannot move. This usually occurs with Black on spot *B*, but it can also occur with Black on spot *4* or *6*. Any other outcome is a win for Black. He wins if he slips behind "enemy lines," making it impossible for White to pin him, or if a situa-

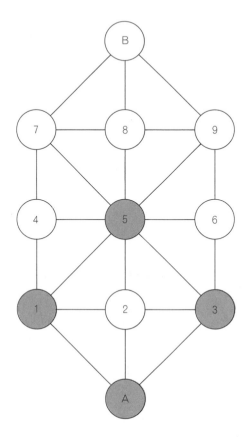

30. The French Military Game

tion develops in which the same moves are endlessly repeated.

The game is as simple to learn as ticktack-toe, but it is more exciting to play and more difficult to analyze. Lucas is able to show that White, if he plays rationally, can always win, but there is no simple strategy and the game abounds in traps and surprises. Often the best move is the move that seems to be the worst. An experienced Black has little

difficulty escaping from an inexperienced White.

Suppose we increase Black's freedom by permitting him to place his piece, at the start of the game, on *any* spot he chooses? Who now wins if both sides play rationally?

Topological board games, on which players construct paths that twist about over the field, are recent developments. Hex, Bridg-it, Zig-Zag, Roadblock, Pathfinder, Squirt, Twixt: these are trade names of some of the games of this type that have been marketed during the past thirty years. In 1960 William L. Black, then an undergraduate at the Massachusetts Institute of Technology, made a study of Hex and Bridg-it, two games discussed in earlier collections of my columns. An outcome of this study was a novel topological game his friends called Black.

Although marked tiles can be used, Black is easily played as a pencil-and-paper game on a checkered field. The size of the field is optional; the standard eight-by-eight field seems ideal, but it is simpler to explain the game on the smaller four-by-four. After the field is drawn the first player starts the game by making a cross in the upper left corner cell as shown in Figure 31. The second player continues the path by making one of three permissible marks in a cell adjacent to the first cell marked. The three marks, shown at the bottom of the illustration, are each composed of two lines. One line represents one of the three ways in which the path can be joined to an open side of the square; the second is added to connect the remaining two sides.

The players alternate moves. Each move

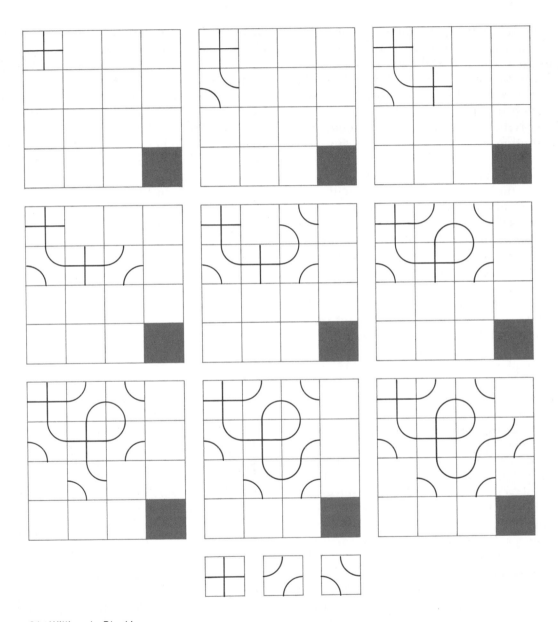

31. William L. Black's game

must extend the path into a neighboring cell. Each player tries to avoid running the path to a border of the field. If he is forced to carry the path to the border, he loses the game. He wins if he succeeds in extending the path into (not just to the border of) the lower right corner cell [*shown shaded*]. The illustration shows successive moves of a typical short game. The first player wins by forcing his opponent to play in the upper right corner cell, where any mark will carry the path to the edge of the field. (Note that the cross extends the path only along one of its arms, although the other arm may become part of the path as the result of a later play.)

The game of Black is of special interest because soon after it was conceived a friend of Black's, Elwyn R. Berlekamp, hit on an elegant strategy that guarantees a win for one of the players. The strategy applies to rectangular fields of any size or shape. Since knowledge of the strategy destroys all interest in actual play, I urge you to play the game and see if you can match Berlekamp's brilliant insight before checking the answer section.

One of the best of many medieval board games is a game that seems to have been first played in Scandinavian countries as early as the fourth and fifth centuries, when it was called *tafl*. In later centuries it was known as *hnefatafl*. The Norsemen introduced the game to Britain, where it was the only board game played by the early Saxons until it began to be replaced by chess in the eleventh and twelfth centuries. H. J. R. Murray, in his *History of Board-Games Other than Chess*, gives reasons for thinking that this is essentially the same game that was still being played in the sixteenth century in Wales, under the name of *tawlbwrdd*, and in the eighteenth century in Lapland, where it was known as *tablut*.

It was Murray who discovered that the great Swedish botanist Carolus Linnaeus included a full description of *tablut* in an extensive diary he kept during his exploration of Lapland in 1732. An English translation of the diary, by Sir James Edward Smith, was published in London in 1811 with the title *Lachesis Lapponica: or a Tour of Lapland.* Figure 32 is a reproduction of the *tablut* board as it is shown on page 55 of Volume II of that edition.

32. The game of tablut

White pieces, representing light-haired Swedes, include a single king and 8 warriors. Black pieces, 16 in number, represent Muscovite warriors. (It is convenient to use a white chess king and 8 white pawns for the Swedes. Black chessmen can be used for the Muscovites, but all must be regarded as identical pieces.) Each black and white piece, including the king, moves like a rook in chess, that is, an unlimited distance along vacant cells in a straight line paralleling a side of the board.

The game begins with the Swedish king occupying the center square, which is known as the castle. Only the king is permitted to stand in the castle, although any piece may move through it when it is vacant. Surrounding the king, on the 8 shaded squares, are his eight warriors. The Muscovites occupy the 16 decorated squares at the four sides of the board.

Either player may open the game. Enemy pieces are captured by a pincer move that consists of occupying adjacent cells on opposite sides of a piece, the three pieces being in the same row or column. For example, if Black makes the indicated move, he captures the three white pieces simultaneously [see Figure 33, top drawing]. If a piece moves between two enemy pieces, however, it is not captured by them. The king may take part in capturing enemy pieces, but he himself is captured only if he is surrounded on all four sides by four enemy pieces or by three enemy pieces and the castle square [middle drawing]; he cannot move from his castle into such a formation without being captured.

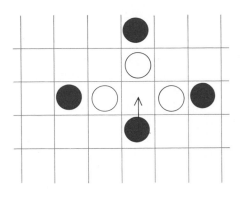

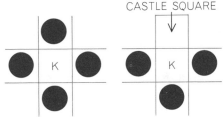

CASTLE SQUARE

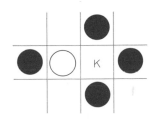

33. Methods of capture in tablut

Linnaeus adds that when the king is in his castle, with three enemy warriors on three sides, and one of his own men on the fourth side, the Swedish warrior is taken if a Muscovite moves to the cell next to the Swede on the side opposite the king [bottom drawing].

Black's objective is to capture the king.

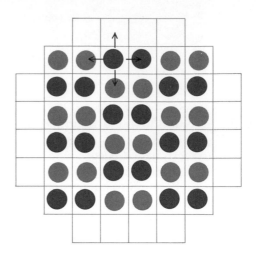

34. Sidney Sackson's game of Focus

If this occurs, the Muscovites win. White's objective is to allow the king to flee the country by reaching any cell on the perimeter of the board. Whenever there is an unobstructed path along a row or column by which the king can reach the border, White must warn Black by saying "Raichi!" (a remark similar in function to "Check!" in chess). If there are two escape paths, White calls out "Tuichu!" Of course "Tuichu!" announces a win for White because there is no way Black can block two escape routes with a single move.

Sidney Sackson, a New York City engineer who makes a hobby of collecting board games, knows of only one occasion on which *tablut* has been made and sold in this country. In 1863 it was issued as a Civil War game called Freedom's Contest, or the Battle for the Union. This game is identical with *tablut* except that the king is called the "Rebel chief" and the pieces are Rebel and Union soldiers. The Rebel chief is limited to a maximum move of four spaces. The traditional game seems to favor White, so perhaps this restriction was intended to redress the balance. (*Breakthru*, Minnesota Mining and Manufacturing's Bookshelf Game currently on sale, is based on *tablut*.)

Sackson is himself the inventor of many unusual board games, one of the best of which he calls Focus. It is played with 36 counters, half of them one color and half another. Small poker chips of the interlocking variety make excellent pieces. They are placed initially on an eight-by-eight board from which three cells at each corner have been removed. Figure 34 shows how the pieces (black and colored in this case) are arranged.

Either side may move first. A move consists of moving a "pile" of pieces (at the outset all piles are one chip high) as many spaces as there are pieces in the pile. Moves are vertical or horizontal, never diagonal. The four possible moves of one colored piece at the start of a game are shown in Figure 34. If the piece moves up, it lands on a vacant square. A move to the right puts it on top of another colored piece, to the left or down puts it on top of a black piece. The last three moves form two-high piles. Such piles may be moved two spaces in any direction. Piles of three, four, and five pieces move three, four, and five spaces respectively. A pile is controlled by the player who owns the piece on top. In moving it does not matter whether the intervening cells are empty or occupied by piles controlled by either player. Passed-over pieces are not affected in any way. A move may end on a vacant cell or on another pile. Figure 35 shows the possible moves of a two-high pile.

Piles may not contain more than five pieces. If a move produces a pile of more than five, all pieces in excess of five are taken from the bottom of the stack. If they are enemy pieces, they are considered captured and are removed from the game. If they belong to the player making the move, they are placed aside as reserves. At any time during the game a player may, if he wishes, take one of his reserve pieces and place it on any cell of the board, empty or otherwise. It has the same effect as a moved piece: if it goes on a pile, the pile belongs to the player who placed it. Using a reserve piece substitutes for a move on the board.

A player may, if he wishes, make a move of fewer spaces than the number of pieces in the pile being moved. He does this by taking from the *top* of the pile as many pieces as the number of spaces he wishes

to move. The rest of the pieces stay where they are. For example, a player may take the top three pieces of a five-high pile and move them three spaces. The pile that *remains* after such a move belongs to the player who owns the piece on top.

When a player is unable to move (that is, controls no piles and has no reserves), the game is over and his opponent wins.

One additional rule is needed. As Paul Yearout, a mathematician at Brigham Young University, pointed out, the second player can always achieve at least a draw by symmetry play; that is, after each move by the first player, he duplicates the move by a symmetrically opposite play. To prevent this, Sackson suggests either of the following alternatives: (1) A draw is declared a win for the first player, (2) Before the game begins each player switches one of his pieces for one of his opponent's pieces (the second player must make an exchange that does not restore symmetry to the pattern) and the game then proceeds as described.

Focus was marketed by Whitman Publishing Company in 1965, the first of Sackson's many marketed games. For a more detailed account of the game as well as suggestions for strategic play, see pages 125–134 of Sackson's *Gamut of Games*.

Answers

Which side wins the French Military Game if Black is given the privilege of starting his piece on any vacant cell? The question was first answered by the Dutch mathematician Frederik Schuh in his book *Wonderlijke*

35. *Moves in the game of Focus*

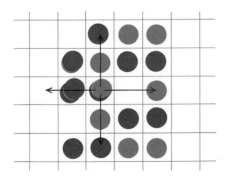

Problemen, published in Holland in 1943. White can always trap Black if he plays rationally. A complete analysis cannot be given here, but the following table shows White's winning responses to Black's six different opening plays.

Black	White
2	A 3 5
4 (or 6)	A 1 5 (or A 3 5)
5	1 2 3
7 (or 9)	A 1 5 (or A 3 5)
8	A 1 5
B	1 2 3

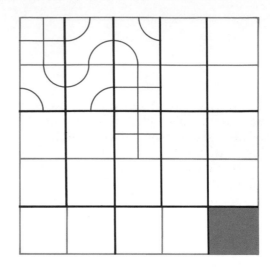

36. Strategy for five-by-five game of Black

For a complete analysis of the game see F. Gobel's translation of Schuh's book, *The Master Book of Mathematical Recreations,* edited by T. H. O'Beirne (New York: Dover, 1968; pages 239–244). Schuh also analyzes variants of the game. For a good suggestion on how to program a computer to play the game see Donald E. Knuth's *Fundamental Algorithms* (New York: Addison-Wesley, 1968; page 546). Richard Sites, a computer scientist at Stanford University, proved in 1970 that White, regardless of where Black starts, can always trap Black on the board's central cell.

The topological game of Black is won on square boards by the first player if the total number of cells is odd, by the second player if the number of cells is even.

When the play is on an odd-celled board, say a five-by-five, the first player's strategy is to suppose the board, except for the lower right corner cell, is completely covered with dominoes [*see Figure 36*]. The way the

dominoes are placed is immaterial. Each move by the second player starts the path on a new domino. The first player then plays so that the path *remains on that domino.* This forces the second player to complete the domino and start the path on another one. It is obvious that the second player eventually will be forced to the border or to an edge of the lower right corner cell.

On even-celled square boards the strategy by which the second player wins is more complicated. The board is thought of as being covered with dominoes except for the upper left and lower right corner cells.

Since the two missing cells are the same color, however (supposing that the board is colored like a checkerboard), it is clearly impossible to cover the remaining cells completely with dominoes: there will always be two uncovered cells of the same color. Elwyn R. Berlekamp, who cracked

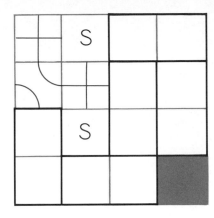

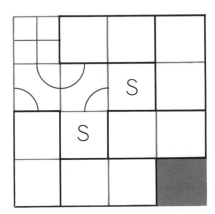

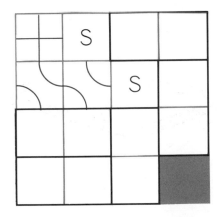

the game, calls these two uncovered cells a "split domino." The split domino is taken care of by the following clever maneuver: The second player makes his first move as shown in Figure 37, top drawing. This forces the first player to play in the second cell of the main diagonal, and his three possible plays are shown. In each case the unused line of his play will connect two cells of the same color. These two cells, labeled *S* in the drawings, are regarded as the split domino. The remaining cells (excluding the lower right corner cell) can now be covered with dominoes. Again, the pattern is arbitrary. The second player wins by the domino method previously explained.

References

Games Ancient and Oriental and How to Play Them. Edward Falkener. London: Longmans, Green and Co., 1892. (Reprint. New York: Dover, 1961.)

A History of Board-Games Other than Chess. H. J. R. Murray. New York: Oxford University Press, 1952.

Board and Table Games from Many Civilizations. R. C. Bell. New York: Oxford University Press, Vol. 1, 1960; Vol. 2, 1969.

A Gamut of Games. Sidney Sackson. New York: Random House, 1969.

6. The Rigid Square and Eight Other Problems

1. The Rigid Square

Raphael M. Robinson, a mathematician at the University of California at Berkeley, is known throughout the world for his solution of a famous minimum problem in set theory. In 1924 Stefan Banach and Alfred Tarski dumfounded their colleagues by showing that a solid ball can be cut into a finite number of point sets that can then be rearranged (without altering their rigid shape) to make two solid balls each the same size as the original. The minimum number of sets required for the "Banach-Tarski paradox" was not established until twenty years later, when Robinson came up with an elegant proof that it was five. (Four are sufficient if one neglects the single point in the center of the ball!)

Here, on a less significant but more recreational level, is an unusual minima problem recently devised by Robinson for which the minimum is not yet known. Imagine that you have before you an unlimited supply of rods all the same length. They can be con-

nected only at their ends. A triangle formed by joining three rods will be rigid but a four-rod square will not: it is easily distorted into other shapes without bending or breaking a rod or detaching the ends. The simplest way to brace the square so that it cannot be deformed is to attach eight more

38. Bracing a square in three dimensions

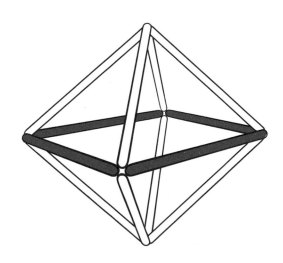

rods [*see Figure 38*] to form the rigid skeleton of a regular octahedron.

Suppose, however, you are confined to the plane. Is there a way to add rods to the square, joining them only at the ends, so that the square is made absolutely rigid? All rods must, of course, lie perfectly flat on the plane. They may not go over or under one another or be bent or broken in any way. The answer is: Yes, the square *can* be made rigid. But what is the smallest number of rods required?

2. A Penny Bet

Bill, a student in mathematics, and his friend John, an English major, usually spun a coin on the bar to see who would pay for each round of beer. One evening Bill said: "Since I've won the last three spins, let me give you a break on the next one. You spin *two* pennies and I'll spin one. If you have more heads than I have, you win. If you don't, I win."

"Gee, thanks," said John.

On previous rounds, when one coin was spun, John's probability of winning was, of course, 1/2. What are his chances under the new arrangement?

3. Three-dimensional Maze

Three-dimensional mazes are something of a rarity. Psychologists occasionally use them for testing animal learning, and from time to time toy manufacturers market them as puzzles. A two-level space maze through which one tried to roll a marble was sold in London in the 1890's; it is depicted in *Puzzles Old and New* by "Professor Hoffmann" (London, 1893). Currently on sale in this country is a cube-shaped, four-level maze of a similar type. Essentially it is a cube of transparent plastic divided by transparent partitions into 64 smaller cubes. By eliminating various sides of the small cubical cells one can create a labyrinth through which a marble can roll. It is a simple maze, easily solved.

Robert Abbott, author of the book *Abbott's New Card Games* (New York: Stein and Day, 1963), recently asked himself: How difficult can a four-by-four-by-four cubical space maze, constructed along such lines, be made? The trickiest design he could achieve is shown in Figure 39. The reader is asked not to make a model but to see how quickly he can run the maze without one.

On each of the four levels shown at the left in the illustration, solid black lines represent side walls. Color indicates a floor; no color, no floor. Hence a small square cell surrounded on all sides by black lines and uncolored is a cubical compartment closed on four sides but open at the bottom. To determine if it is open or closed at the top it is necessary to check the corresponding cell on the next level above. The top level (A) is of course completely covered by a ceiling.

Think of diagrams A through D as floor plans of the four-level cubical structure shown at the right in the illustration. First see if you can find a path that leads from the

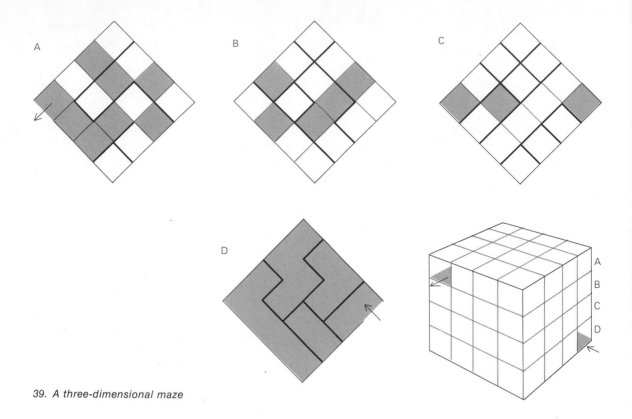

39. A three-dimensional maze

entrance on the first level to the exit on the top level. Then see if you can determine the shortest path from the entrance to the exit.

4. Gold Links

Lenox R. Lohr, president of the Museum of Science and Industry in Chicago, was kind enough to pass along the following deceptively simple version of a type of combinatorial problem that turns up in many fields of applied mathematics. A traveler finds himself in a strange town without funds; he expects a large check to arrive in a few weeks. His most valuable possession is a gold watch chain of 23 links. To pay for a room he arranges with a landlady to give her as collateral one link a day for 23 days.

Naturally the traveler wants to damage his watch chain as little as possible. Instead of giving the landlady a separate link each day he can give her one link the first day, then on the second day take back the link and give her a chain of two links. On the third day he can give her the single link again and on the fourth take back all she has and give her a chain of four links. All that matters is that each day she must be in possession of a number of links that corresponds to the number of days.

The traveler soon realizes that this can be accomplished by cutting the chain in many different ways. The problem is: What is the smallest number of links the traveler needs to cut in order to carry out his agreement for the full 23 days? More advanced mathematicians may wish to obtain a general formula for the longest chain that can be used in this manner after n cuts are made at the optimum places.

5. Word Squares

Word puzzlists have long been fascinated by a type of puzzle called the word square. The best way to explain this is to provide an example:

```
M E R G E R S
E T E R N A L
R E G A T T A
G R A V I T Y
E N T I T L E
R A T T L E R
S L A Y E R S
```

Note that each word in the above order-7 square appears both horizontally and vertically. The higher the order, the more difficult it is to devise such squares. Word square experts have succeeded in forming many elegant order-9 squares, but no order-10 squares have been constructed in English without the use of unusual double words such as Pango-Pango.

Charles Babbage, the nineteenth-century pioneer in the design of computers, explains how to form word squares in his autobiography, *Passages from the Life of a Philosopher,* and adds: "The various ranks of the church are easily squared; but it is stated, I know not on what authority, that no one has succeeded in squaring a bishop." Readers of *Eureka,* a mathematics journal published by students at the University of Cambridge, had no difficulty squaring *bishop* when they were told of Babbage's remarks. The square shown below (from the magazine's October 1961 issue) was one of many good solutions received:

```
B I S H O P
I L L U M E
S L I D E S
H U D D L E
O M E L E T
P E S E T A
```

As far as I know, no one has yet succeeded—perhaps even attempted—to square the word "circle." Only words found in an unabridged English dictionary may be used. The more familiar the words, the more praiseworthy the square.

6. The Three Watch Hands

Assume an idealized, perfectly running watch with a sweep second hand. At noon all three hands point to exactly the same spot on the dial. What is the next time at which the three hands will be in line again, all pointing in the same direction? The answer is: Midnight.

The first part of this problem—much the easiest—is to prove that the three hands are together only when they point straight up. The second part, calling for more ingenuity, is to find the exact time or times, between noon and midnight, when the three hands come *closest* to pointing in the same direction. "Closest" is defined as follows: two hands point to the same spot on the dial, with the third hand a minimum distance away. When does this occur? How far away is the third hand?

It is assumed (as is customary in problems of this type) that all three hands move at a steady rate, so that time can be registered to any desired degree of accuracy.

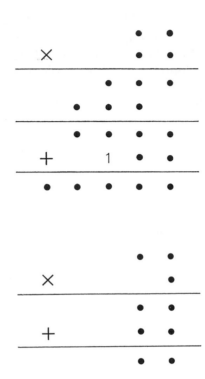

7. Three Cryptarithms

Of the three remarkable cryptarithms in Figure 40 the first [*top*] is easy, the second [*middle*] is moderately hard, and the third [*bottom*] is so difficult that I do not expect any reader to solve it without the use of a computer.

Problem 1: Each dot represents one of the ten digits from 0 to 9 inclusive. Some digits may appear more than once, others not at all. As you can see, a two-digit number multiplied by a two-digit number yields a four-digit product, to which is added a three-digit number starting with 1. Replace

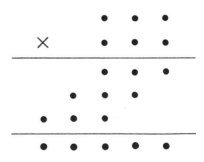

40. *Three cryptarithms*

each dot with the proper digit. The solution is unique.

Problem 2: As in the first cryptarithm, a multiplication is followed by an addition. In this case, however, each dot is a digit from 1 to 9 inclusive (no 0) and each digit appears once. The answer is unique.

Problem 3: Each dot in this multiplication problem stands for a digit from 0 to 9 inclusive. Each digit appears exactly *twice*. Again, the answer is unique.

8. Maximizing Chess Moves

When the eight chess pieces of one color (pawns excluded) are placed alone on the board in the standard starting position, 51 different moves can be made. Rooks and bishops can each make 7 different moves, knights and the king can each make 3, the queen can make 14. By changing the positions of the pieces it is easy to increase the number of possible moves. What is the maximum? In other words, how can the

eight pieces of one color be placed on an empty board in such a way that the largest possible number of different moves can be made?

The two bishops should be placed on opposite color squares to conform with standard chess practice, and the move of castling is not considered. Actually neither qualification is necessary because in both cases a violation would only restrict the freedom of pieces to move.

9. Folding a Möbius Strip

Stephen Barr's method of folding a Möbius strip from a square sheet of paper was explained in chapter 2. The square [*at left in Figure 41*] is simply folded in half twice along the dotted lines, then edge B is taped to B'. The result is a band with a half-twist, one-sided and one-edged; it is a legitimate model of a Möbius surface even though it cannot be opened out for easy inspection.

Suppose instead of a square we use a paper rectangle twice as long as it is high

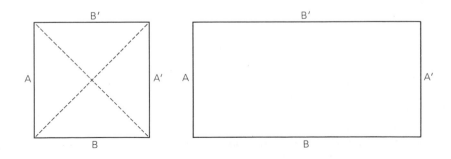

41. Möbius-strip problem

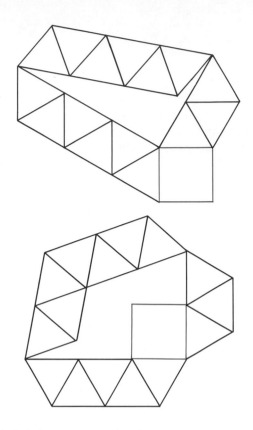

[*at right in Figure 41*]. Is it possible to fold *this* into a Möbius surface that joins B to B'? One can fold or twist the paper in any way, but of course it must not be torn. Assume that the paper can be made as thin as desired. The surface must be given a half-twist that allows the entire length of edge B to be joined to the entire length of edge B'. It would not be difficult to make the strip by joining A to A'; the problem is to find a way to do it by connecting the pair of longer edges.

Once the reader has either found a way to do it or concluded that it is impossible, a more interesting question arises: What is the smallest value for A/B that will allow a Möbius strip to be folded by the joining of B to B'?

Answers

42. Solutions to square-bracing problem

1.

Raphael M. Robinson's best solution to his problem of bracing a square on the plane with the minimum number of rods, all equal in length to the square's side, calls for 31 rods in addition to the 4 used for the square. Figure 42 shows two of several equally good patterns.

This answer was reduced to 25 rods [*see Figure 43*] by 57 *Scientific American* readers. As I was recovering from the shock of this elegant improvement seven readers — G. C. Baker, Joseph H. Engel, Kenneth J. Fawcett, Richard Jenney, Frederick R. Kling, Bernard M. Schwartz, and Glenwood

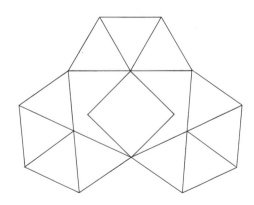

43. 25-rod solution for square-bracing

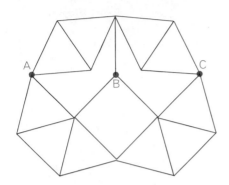

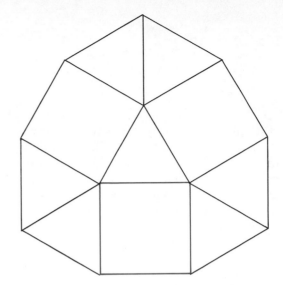

44. 23-rod solution

Weinert — staggered me with the 23-rod solution shown in Figure 44. Later, about a dozen more readers sent the same solution. The rigidity of the structure becomes apparent when one realizes that points *A*, *B*, and *C* must be collinear.

All solutions with fewer than 23 rods proved to be either nonrigid or geometrically inexact. For example, many readers sent the pattern shown in Figure 45, top, which is not rigid, or the pattern shown at the bottom, which, although rigid, unfortunately includes line *A*, a trifle longer than one unit.

The problem obviously can be extended to other regular polygons. The hexagon is simply solved with internal braces (how many?), but the pentagon is a tough one. T. H. O'Beirne managed to rigidify a regular pentagon with 64 additional rods, but it is not known if this number is minimal.

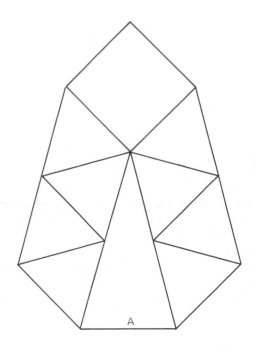

45. Two incorrect solutions to square-bracing

2.

Bill spins one penny, John two. John wins if he has more heads than Bill. A tabulation of the eight equally probable ways 3 coins can fall shows that John wins in four cases and loses in four, so his chances of winning are 1/2, which is what they would be if a single coin were spun. His probability of winning remains the same whenever he has one more coin than Bill. Thus if he has 51 coins and Bill has 50, each man still has an equal chance of winning. This problem appeared in the Canadian magic magazine *Ibidem;* December, 1961, page 24; it was contributed by "Ravelli," pen-name of chemist Ronald Wohl.

3.

The simplest paper-and-pencil way to solve Robert Abbott's three-dimensional maze is to place a spot in each cell and then draw lines from spot to spot to represent all open corridors. Since a maze involves only topological properties of the pattern, it does not matter how these lines twist and turn as long as they connect the spots properly.

The next step is to erase all blind-alley lines and all loops that do no more than take one from spot to spot in a roundabout way when a shorter path is available. Eventually only the shortest route remains. This path is shown in Figure 46. Note that two loops

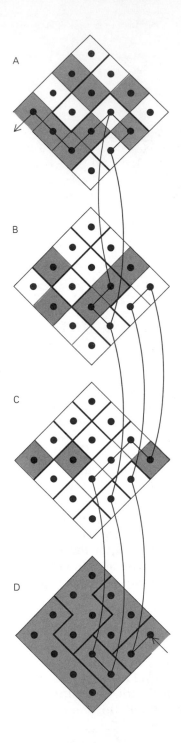

46. Three-dimensional maze solution

near the top offer two different routes of equal length. Each of the long curved lines connecting the spots is, of course, only one unit in length in the actual maze, therefore the entire maze can be run by a path 19 units long.

An alternate method of finding the shortest path in any type of maze is to make a model of the network out of string. Each segment of string must have a length that is in the same proportion to the length of the corridor it represents, and it must be labeled in some way so that the corridor can be identified. After the model is completed, pick up the "start" of the network with one thumb and finger and the "end" of the network with the other thumb and finger. Pull the string taut. Roundabout loops and blind alleys hang loose. The taut portion of the model traces the path of minimum length!

A third method is to label the starting cell with "1". Put "2" in all cells that can be reached in one step. Put "3" in all cells that can be reached in one step from each 2-cell. Continue in this manner, numbering every cell once. If you return to a cell already labeled, do not give it a larger number. After all cells are labeled, start at the final cell and move backward through the numbered cells, taking them in reverse order, to trace out a minimal-length path.

There is now a large literature on these and other algorithms for finding shortest routes in mazes or on graphs. Recent references follow; readers will find many earlier articles on the topic listed in them:

"The Shortest Path through a Maze." E. F. Moore. In *Proceedings of an International Symposium on the Theory of Switching*, Part II, April 2–5, 1957. (Reprinted in *Annals of the Computation Laboratory of Harvard University*, Vol. 30, 1959. Pages 285–292.)

"An Algorithm for Path Connections and Its Applications." C. Y. Lee. *I. R. E. Transactions on Electronic Computers*, Vol. EC–10; September, 1961. Pages 346–365.

"All Shortest Routes in a Graph." G. B. Dantzig. *Operations Research Technical Report 66–3*, Stanford University; November, 1966.

"Shortcut in the Decomposition Algorithm for Shortest Paths in a Network." T. C. Hu and W. T. Torres. *IBM Journal of Research and Development*, Vol. 13, No. 4; July, 1969. Pages 387–390.

Algorithms, Graphs, and Computers. Richard Bellman, Kenneth L. Cooke, and Jo Ann Lockett. New York: Academic Press, 1970. Pages 94–100.

4.

The traveler with a 23-link gold chain can give his landlady one link a day for 23 days if he cuts as few as 2 links of the chain. By cutting the fourth and eleventh links he obtains two segments containing one link and segments of 3, 6, and 12 links. Combining these segments in various ways will make a set of any number of lengths from 1 to 23.

The formula for the maximum length of chain that can be handled in this way with n cuts is

$$[(n + 1)2^{n+1}] - 1.$$

Thus one cut (link 3) is sufficient for a chain

of 7 links, three cuts (links 5, 14, 31) for a chain of 63 links, and so on.

5.

I confess that what I thought was a new problem turns out, as Dmitri Borgmann informed me, to be one of the first English word squares ever published! In a letter to the British periodical *Notes and Queries* for July 21, 1859, a reader signing himself "W. W." spoke of the word-squaring game "which has of late been current in society" and proceeded to give the following example: Circle, Icarus, Rarest, Create, Lustre, Esteem. "There are very probably," he wrote, "other ways of squaring the circle."

Yes, when I published the problem in *Scientific American* about 1,000 readers found more than 250 different ways of doing it. I despair of summarizing the variations. The most popular choice for a second word was Inures, with Iberia, Icarus, and Isohel following in that order. The square composed by the most (227) people was: Circle, Inures, Rudest, Crease, Lesser (or Lessor), Esters. Almost as many (210) sent essentially the same square, with Lessee and Esteem as the last two words. "This was done with *ease*," wrote Allan Abrahamse, in punning reference to the fact that a main diagonal of this square consists entirely of *E*'s. Fifty-six readers found Circle, Inures, Rumens, Create, Lenten, Essene.

The most popular square with Iberia as the second word was Circle, Iberia, Recent (or Relent, Repent, and so on), Create, Linter, Eaters (or Eatery). The most popular with Icarus second: Circle, Icarus, Rarest, Create, Luster, Esters. With Isohel second: Circle, Isohel, Roband (or Roland), Chaise (or Chasse), Lenses, Eldest. Each of these three squares was arrived at by more than a hundred readers.

Of some 40 other words chosen for the second spot, Imaret was the favorite. More than 40 readers used it, mostly as follows: Circle, Imaret, Radish, Crissa, Lesson, Ethane. Many squares with unusual words were found by one reader only; the following are representative:

Circle, Imoros, Romist, Crimea, Losest, Estate (Frederick Chait).

Circle, Isolux, Rosace, Claver, Lucent, Exerts (Ross and Otis Schuart).

Circle, Iterum, Refine, Cringe, Lunger, Emeers (Ralph Hinrichs).

Circle, Isaian, Rained, Cingle, Laelia, Endear (Robert Utter).

Circle, Ironer, Rowena, Cnemis, Lenite, Eraser (Ralph Beaman).

Circle, Inhaul, Rhymed, Camise, Lueses, Eldest (Riley Hampton).

Circle, Irenic, Regime, Cnidus, Limuli, Ecesis (Mrs. Barbara B. Pepelko).

A number of readers tried the more difficult task of squaring the square. All together about 24 different squared squares came in, all with esoteric words such as Square, Quaver, Uakari, Avalon, Rerose, Erinea (Mrs. P. J. Federico). Several readers tried to square the triangle, but without success. Edna Lalande squared the ellipse: Ellipse, Lienees, Lecamas, Inagile, Pemican, Sealane, Essenes.

Four readers (Quentin Derkletterer,

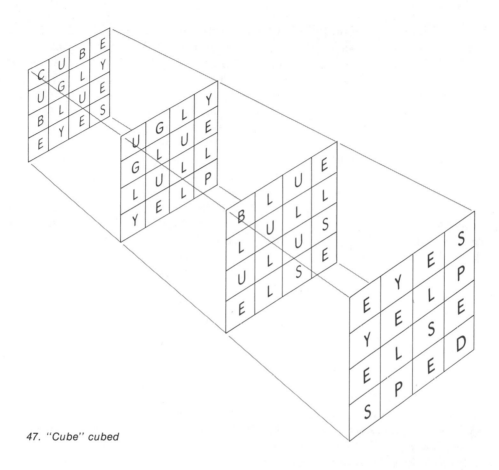

47. "Cube" cubed

Solomon Golomb, John McClellan, and James Topp) independently hit on this delightful squared cube:

```
C U B E
U G L Y
B L U E
E Y E S
```

Derkletterer took off from this square, along a third co-ordinate, and managed to cube the cube: Cube, Ugly, Blue, Eyes; Ugly, Glue, Lull, Yelp; Blue, Lull, Ulus, Else; Eyes, Yelp, Else, Sped [*see Figure 47*]. Patrick O'Neil and Charles Keith cubed the cube this way: Cube, Upon, Bold, Ends; Upon, Pole, Olio, Neon; Bold, Olio, Liar, Dora; Ends, Neon, Dora, Snap. Benjamin

F. Melkun and Glenn A. Larson found still another cubed cube, then vanished along a fourth co-ordinate and came back with hypercubes for the words Pet and Eat. They were unable to cube the sphere. R. J. Rea was able to cube eggs; and H. M. Thomas and H. P. Thomas cubed root, dice, beef, and ice, but were unable to cube sugar.

Leigh Mercer, the London expert on word play, sent me the best-known squares in which the words, taken in order, form sentences:

> Just, Ugly, Slip, Type.
> Might, Idler, Glide, Hedge, Trees?
> Crest, Reach, Eager (Scene Three).
> Leave, Ellen, Alone, Venom, Enemy.

6.

A quick way to prove that all three hands of a watch with a sweep second hand are together only when they point to 12 is to apply elementary Diophantine analysis. When the hour hand coincides with one of the other hands, the difference between the distances traveled by each must be an integral number of hours. During the 12-hour period the hour hand makes one circuit around the dial. Assume that it travels a distance x, less than one complete circuit, to arrive at a position with all three hands together. After the hour hand has gone a distance of x, the minute hand will have gone a distance of $12x$, making the difference $11x$. In the same period of time the second hand will have gone a distance of $720x$, making the difference $719x$. All three hands can be together only when x has a value that makes both $719x$ and $11x$ integral. But 719 and 11 are both prime numbers, therefore x can take only the values of 0 and 1, which it has at 0 and 12 o'clock respectively.

Aside from the case in which all three hands point straight up, the closest the hands come to pointing in the same direction (defining "closest" as the minimum deviation of one hand when the other two coincide) is at 16 minutes 16 and 256/719 seconds past 3, and again at 43 minutes 43 and 463/719 seconds after 8.

The two times are mirror images in the sense that if a watch showing one time is held up to a mirror and the image is read as though it were an unreversed clock, the image would indicate the other time. The sum of the two times is 12 hours. In both instances the second and hour hands coincide, with the minute hand separated from them by a distance of 360/719 of one degree of arc. (The distance is 5 and 5/719 seconds if we define a second as a sixtieth of the distance of a clock minute.) In the first instance the minute hand is behind the other two by this distance, in the second instance it leads by the same distance.

Another simple proof that the three hands are never together except at 12 was found by Henry D. Friedman, of Sylvania Electronic Systems. The hour and minute hands meet eleven times, with periods of 12/11 hours that divide the clock's circumference into eleven equal parts. The minute and second hands similarly divide the circumference into 59 equal parts. All three hands can meet only at a point where $r/11 = s/59$, r and s being positive integers with r less than 11 and s less than 59. Since 11/59 cannot be reduced to a lower fraction, r/s, there can be no meeting of the three hands except at 12 o'clock.

7.

The three cryptarithms have the unique solutions shown in Figure 48. The top one

was devised by Stephen Barr, the lower left one is the work of the English puzzlist Henry Ernest Dudeney; the lower right one is from Frederik Schuh's *Wonderlijke Problemen*. For a translation of Schuh's analysis see pages 287–291 in F. Gobel's *Master Book of Mathematical Recreations*.

The third cryptarithm, which I thought no one could solve without a computer, was solved with pencil and paper by no fewer than 53 *Scientific American* readers. Few of the solvers went on to show that no other

```
          9 9
   ×      9 9
       8 9 1
     8 9 1
     9 8 0 1
   + 1 9 9
   1 0 0 0 0
```

48. Solutions to cryptarithms

```
     1 7
   ×   4
     6 8
   + 2 5
     9 3
```

```
         1 7 9
   ×     2 2 4
         7 1 6
       3 5 8
     3 5 8
     4 0 0 9 6
```

solution was possible. Six readers, however, programed computers to check all possibilities, and they confirmed the uniqueness of the answer.

8.

If eight chess pieces of one color are placed on the board as shown in Figure 49, a total of exactly 100 different moves can be made. According to T. R. Dawson, the English chess problemist, this question was first asked in 1848 by a German chess expert, M. Bezzel. His solution, the one shown here, was published the following year. In 1899 E. Landau, in *Der Schachfreund*, September, 1899, proved that 100 moves is the maximum and that Bezzel's solution is unique except for the trivial fact that the rook, on the seventh square of the fourth row from the top, could just as well be placed on the first square of that same row.

Among the many readers who solved this chess problem, fourteen supplied detailed proof that 100 moves is indeed the maximum.

For a way of placing the eight pieces so that a *minimum* number of moves (ten) are possible, see Figure 38, page 88, of my book *Unexpected Hanging*.

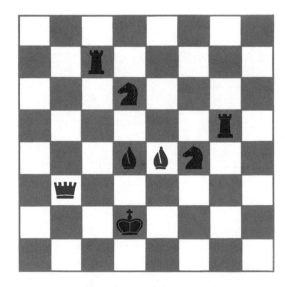

49. Board setup for maximum chess moves

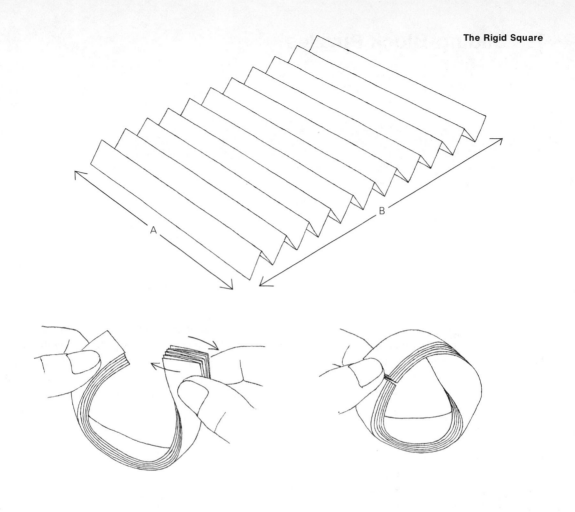

50. Möbius-strip solution

9.

What is the smallest value of *A/B* that allows one to join the *B* edges of the paper rectangle into a Möbius band? The surprising answer is that there *is* no minimum. The fraction *A/B* can be made as small as one pleases.

Proof is supplied by a folding technique explained in Stephen Barr's *Experiments in Topology* (New York: T. Y. Crowell, 1964). The strip is pleated as shown in Figure 50 to form a narrow strip with ends that show two-fold symmetry. After this narrow strip is given a half-twist the ends are joined. *Voilà!*

7. Sliding-Block Puzzles

"THE OLDER INHABITANTS of Puzzleland," wrote Sam Loyd in his *Cyclopedia of Puzzles,* "will remember how in the early seventies I drove the entire world crazy over a little box of movable blocks which became known as the 14–15 Puzzle." Fifteen numbered blocks were placed in a square box as shown in Figure 51. The object was to slide the blocks about, one at a time, until the 14–15 error was corrected and all blocks were in serial order with the empty space in the lower right-hand corner as before.

The craze spread rapidly to Britain and Europe. "People became infatuated with the puzzle," Loyd continued, "and ludicrous tales are told of shopkeepers who neglected to open their stores; of a distinguished clergyman who stood under a street lamp all through a wintry night trying to recall the way he had performed the feat. . . . A famous Baltimore editor tells how he went for his noon lunch and was discovered by his frantic staff long past midnight pushing little pieces of pie around on a plate!"

Interest in the puzzle abated after several mathematicians published articles proving it could not be done. Today the puzzle (still on sale in a variety of forms) is sometimes cited by computer experts as a miniature model of what is now called a sequential

51. *Sam Loyd's 14–15 Puzzle*

machine. Each movement of a block is an input, each arrangement, or "state," of the blocks is an output. It turns out that exactly half of the 15! ($1 \times 2 \times 3 \ldots \times 15$), or 1,307,674,368,000, possible states of the machine are achievable outputs. The mathematical theory of the 14–15 Puzzle applies to all sliding-block puzzles in which the pieces are unit squares confined to rectangular fields.

But not to sliding-block puzzles in which the pieces are *not* unit squares! The success of Loyd's puzzle brought a rash of sliding-block puzzles, with differently shaped pieces, that have sold all over the world for the past eighty years. These puzzles are very much in want of a theory. Short of trial and error, no one knows how to determine if a given state is obtainable from another given state, and if it is obtainable, no one knows how to find the minimum chain of moves for achieving the desired state. These entertaining puzzles provide all sorts of challenges for computer programmers. For the rest of us they are engrossing solitaire games that can be constructed in a few minutes with only a pair of scissors and a supply of cardboard.

A puzzle of this type — perhaps the earliest and certainly the most widely sold — is shown in Figure 52. The reader is urged to stop reading and cut the nine pieces from a sheet of thin cardboard. The diagram is easily copied by drawing a four-by-five rectangle, ruling it lightly into unit squares, then outlining the nine pieces. Number them as indicated, cut them out and place them on a four-by-five rectangle drawn on

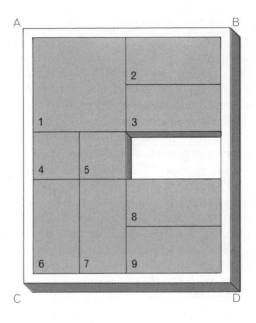

52. *Dad's Puzzle*

a sheet of paper or cardboard of contrasting color. The problem: By sliding the pieces one at a time, keeping them flat on the paper and inside the rectangle, bring the large square from corner A to corner C.

It is easy to bring the square to corner B. Move the pieces in order as follows: 5, 4, 1, 2, 3; 4 (up and right), 1, 6, 7, 8; 9, 5, 4, 1, 6; 7 (halfway), 9, 5, 4, 8; 6, 2, 3, 1. This is a minimum-move solution in 24 steps for which I am indebted to Edward E. Roderick, Alfred C. Collins, Jan-Henrik Johansson, and Michel Hénon. (Sliding a piece "around a corner" is counted as one move.) To bring the large square to corner *D* requires 29 moves. The first 19 are the same as before, then continue with: 1, 3, 2, 6, 7; 8, 9, 4, 5, 1.

It is not possible to slide the large square from corner *A* to corner *C* in fewer than 59 moves. Readers are urged to see if they can achieve this minimum before the moves are disclosed. Cardboard pieces are quite satisfactory, although handsomer and more permanent models can be cut from sheets of wood, plastic, linoleum, Vinylite, and so on. The restraining border can be made by gluing strips on a wooden board. The board should be sandpapered for smooth sliding, and it is best to round off the corners of the pieces and bevel their edges slightly.

The origin of this excellent puzzle is unknown. The earliest version in the puzzle collection of the late Lester Grimes of New Rochelle, New York, is called the Pennant Puzzle and was copyrighted in 1909 by L. W. Hardy and made by the O.K. Novelty Company in Chicago. Cardboard pieces bear the names of major cities. The large square, which represents the home team, is to be brought to the corner, which symbolizes first place in the league. In 1926 a wooden version was marketed under the name of Dad's Puzzler, and most later versions have been called Dad's Puzzle. An inexpensive version currently on sale has the trade name Moving Day Puzzle (a picture of a piano is on the large square), and there is an elegant version called Magnetic Square Puzzle with large wooden pieces (containing magnets) that cling to a metal field.

If one of the two-by-one rectangles in

Dad's Puzzle is cut in half to make two unit squares, the resulting ten pieces provide the sliding blocks for a more difficult puzzle [*see Figure* 53] that has long been popular in France under the name of L'Âne Rouge (The Red Donkey). The object of the puzzle is to move the large square with the red donkey's picture to the bottom of the border so that it can be slid out of the box through the opening. A correspondent in Scotland recalls seeing an English version on sale in the early 1930's. More recently it has been sold in this country under such trade names as Intrigue, Mov-it Puzzle and Hako. The minimum-move solution re-

53. *L'Ane Rouge puzzle*

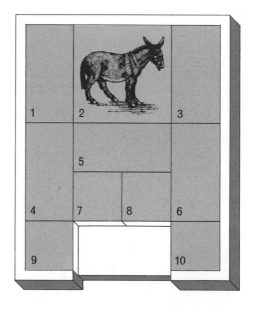

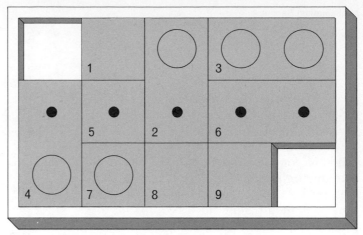

START

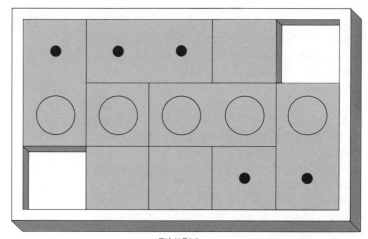

54. Line Up the Quinties puzzle

FINISH

quires 81 moves. It was worked out by Thomas B. Lemann, a New Orleans attorney; and it was proved minimal in 1964 by John Larmouth of Cambridge University, and later by Michel Hénon, both men using computers.

In 1934, when the Dionne quintuplets were born, the event was celebrated by the appearance of an unusual sliding-block puzzle called Line Up the Quinties. (The box bears the imprint of the Embossing Company of Albany, New York, and states that the puzzle was created by Richard W. Fatiguant.) In the schematic drawings of this puzzle [*Figure 54*] the five circles are the faces of the five quintuplets. The problem is to start with the pieces arranged as shown in the first drawing and move them to the pattern shown at the bottom. A 30-move solution, the best I have found, is given in the answer section.

It was inevitable that someone would think of complicating this sort of puzzle by introducing nonrectangular pieces. In 1927 Charles L. A. Diamond of Newburgh, New York, obtained patent No. 1,633,397 for the puzzle shown in Figure 55. It was manufactured under the name of Ma's Puzzle (in obvious competition with Dad's) by the Standard Trailer Company of Cambridge Springs, Pennsylvania. Piece No. 2 was labeled "Ma," No. 5 "My Boy." (The other seven pieces bore the labels "No Work," "Danger," "Broke," "Worry," "Trouble," "Homesick," and "Ill.") The object of the puzzle is to unite Ma with My Boy to form a single three-by-two rectangle in the upper right-hand corner of the box. (This rectangle may be either wider than high or vice versa.) I give a 23-move solution in the answer section. More complicated puzzles, some all rectangular, others with L-shaped pieces, have been marketed here and abroad. Sliding block puzzles with triangular pieces have been explored, chiefly by T. H. O'Beirne of Glasgow, but none have so far been manufactured.

The latest innovation in this curious and unchronicled field has been supplied by Sherley Ellis Stotts, a piano tuner who lives in Denver. Stotts, who holds a master's degree in psychology from the University of Colorado (his thesis was on the reliability of the Seashore music tests), has been blind since the age of seven. In recent years he has invented and made a

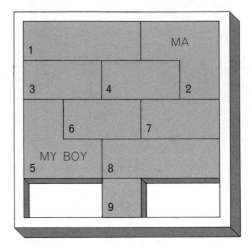

55. Ma's Puzzle

variety of unusual puzzles out of wire, wood and plastic. A patent application is now pending for what he calls his Tiger series of sliding-block puzzles.

Each tiger puzzle is based on a diagram often used by algebra teachers as a visual display of the square of a polynomial. I shall describe only the simplest Tiger puzzle, which exploits the diagram [*Figure* 56] for the square of $a + b + c$. The three terms are represented by the horizontal and vertical line segments on the sides of the square. When the expression is multiplied by itself, the result is $a^2 + b^2 + c^2 + 2ab + 2ac + 2bc$. Each term, of course, is represented in the figure: there are three squares with sides, respectively, of a, b, and c, two rectangles with sides ab, two with sides ac, and two with sides bc. Stotts converted this dissection to the charming puzzle shown at

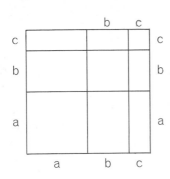

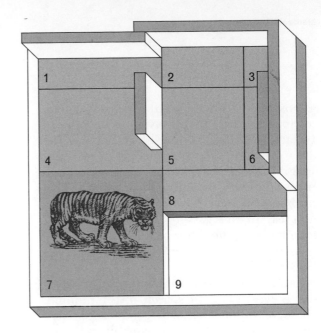

56. Stotts's Tiger puzzle

the right in the illustration. On the large square he glued a replica of a tiger. At the upper right-hand corner he attached to the frame two segments of a fence (shown in color). Three other fence segments were glued to pieces 1, 4, and 6 as shown. (Readers who wish to try the puzzle may simply draw the fences on cardboard pieces.) The ratios of *a:b:c* must be 3:2:1.

The puzzle starts with the pieces arranged as indicated, except that piece 9 is removed from the field. The problem is to slide the pieces so that the tiger square is moved to the upper right-hand corner and completely surrounded by a square fence. Unlike all previous sliding-block puzzles, the open space is large enough to allow, at times, the 90-degree *rotation* of a rectangular piece. This is permitted, of course, only when the rotation is geometrically possible

within the space, keeping all pieces flat on the field. In the answer section I give a 48-move solution.

At the moment there is no practical application for a theory of sliding-block puzzles with differently shaped pieces, but it would be foolhardy to say that none will ever be found. As automation advances, complex problems arise in connection with the efficient storage and retrieval of goods. The day may come when a housewife will dial an order to a department store and machines will find the items and deliver them to a post office or truck. If the items are kept in rectangular packages, it is not inconceivable that a certain amount of package-shifting, within confined areas, will be called for. Something of this sort actually goes on constantly in big-city garages and parking lots where it is necessary to park as many cars as possible within the available space and to retrieve the cars with maximum efficiency. In fact, in Britain sliding-block puzzles are often called "garage puzzles" because several British versions have presented the pieces as cars confined to a garage. The problem, of course, is to maneuver a certain car to the garage's entrance without taking any of the other cars outside.

As the reader will quickly discover if he tries to solve any of these puzzles, there is an almost hypnotic fascination in pushing the pieces about in search of a minimum chain of inputs that will produce the desired state. It is by no means all trial and error. The mind soon "sees" that certain lines of play lead to blind alleys whereas other lines of play are promising.

Answers

Dad's Puzzle: 59 moves. 5, 4, 1, 2, 3, 4 (up, right). 1, 6, 7, 8, 9, 5. 4, 1, 6, 7, 8, 9. 5 (left, up), 9, 8, 5, 4, 1. 3, 2, 7, 6, 4 (up, left), 6. 7, 4, 5, 6, 7, 5 (right, up). 3, 2, 5, 4, 3, 2. 4 (down, right), 2, 3, 6, 7, 1. 4, 5, 2, 3, 6, 7. 1, 4 (left, up), 9, 8, 1.

L'Âne Rouge: 81 moves. 9 (halfway), 4, 5, 8 (down), 6. 10 (halfway), 8, 6, 5, 7 (up, left). 9, 6, 10 (left, down), 5, 9. 7, 4, 6, 10, 8. 5, 7 (down, right), 6, 4, 1. 2, 3, 9, 7, 6. 3, 2, 1, 4, 8. 10 (right, up), 5, 3, 6, 8. 2, 9, 7 (up, left), 8, 6. 3, 10 (right, down), 2, 9, (down, right), 1. 4, 2, 9, 7 (halfway), 8. 6, 3, 10, 9 (down), 2. 4, 1, 8, 7, 6. 3, 2, 7, 8, 1.

4, 7 (left, up), 5, 9, 10. 2, 8, 7, 5, 10 (up, left). 2.

Line Up the Quinties: 30 moves. 9, 8, 1, 2, 3. 6, 8 (up, left), 2, 5 (right, down), 3. 6, 8 (up, left), 9, 2, 8. 6, 3, 1 (right, down), 6, 3. 5 (up, right), 1 (right, down), 7, 1 (left), 8. 5 (down), 3, 6 (halfway), 4, 9.

Ma's Puzzle: the 32-move solution I originally published in the pages of *Scientific American* was reduced by more than a dozen readers to 23. 9 (left), 8, 7, 6, 5. 9 (up), 8, 7, 6, 4. 2, 1, 3 (up), 9 (up, right), 5 (left, up). 6, 4 (down, right), 9 (down all the way), 5, 3. 1, 2, 5.

Tiger Puzzle: 48 moves. Letters stand for up, down, left, right, and turn (90 degrees). 8d, 5d, 6d, 4r, 1d. 2l, 3l, 4u, 1r, 7u. 8l, 5d, 6d, 1d, 4d. 3r, 2r, 7u, 1l, 5u. 6u, 8r, 1d, 5l, 6l. 4d, 2dtr, 7r, 5u, 6l. 4l, 2dl, 3d, 7r, 5r. 6u, 4lu, 1u, 8l, 3d. 2td, 1ru, 8ur, 4d, 6dr. 5l, 6u, 4u. This solution, with one less than the number of moves in the solution I originally presented, was provided by Charles Clapham, John Harris, and Thomas Kew.

References

Mathematical Recreations and Essays. W. W. Rouse Ball. New York: The Macmillan Co., 1960. Pages 299–303.

Mathematical Puzzles and Other Brain Twisters. Anthony S. Filipiak. New York: A. S. Barnes and Co., 1964. Pages 1–18.

8.   Parity Checks

"WHY FOUR KISSES, you will say . . ." wrote John Keats in a letter, commenting on the above stanza from his well-known poem *La Belle Dame sans Merci*. "I was obliged to choose an even number that both eyes might have fair play. . . . I think two a piece quite sufficient. Suppose I had said seven; there would have been three and a half a piece — a very awkward affair."

If we had been told that Keats's pale knight kissed the lady's eyes 37 times, would it be necessary to make an empirical test to determine if each eye could receive the same number of kisses? No, 37 is an odd number, not evenly divisible by 2. We know at once that one eye must have been kissed at least one more time than the other.

An old joke along similar lines tells of a graduate student in mathematics who was on a spring outing with his girl. She plucked a daisy and began to pull off the petals while she recited "He loves me, he loves me not . . . "

You are really going to a great deal of unnecessary trouble," said the young man. "All you have to do is count the petals. If the total is even, the answer is negative. If it is odd, the answer is affirmative."

We have here two trivial applications of what mathematicians sometimes call a parity check. It is one of the most powerful tools in mathematics. Whenever a problem involves odd and even, or two mutually exclusive sets that can be identified with odd and even numbers, a parity check often furnishes a quick, elegant proof for something that might otherwise be extremely difficult to establish.

The classic instance in number theory is provided by Euclid's proof, which may go back to the Pythagoreans, that the square root of 2 cannot be expressed as a common fraction (a fraction with an integer above and an integer below the line). Since the diagonal of a unit square has a length equal to the square root of 2, this means that no ruler, however finely graduated, that accurately measures the side of the square will accurately measure the diagonal.

The proof is easy to follow. Assume that there *is* such a common fraction, n/m, which has been reduced to its lowest terms. Since the square of this fraction is 2, we can write the equation

$$2 = n^2/m^2 \qquad (1)$$

and then rearrange the terms to

$$n^2 = 2m^2 . \qquad (2)$$

The right side of this equality is an even number (because it is a multiple of 2); therefore the left side, n^2, is even. Only an even number gives an even product when multiplied by itself; therefore n also is even. We turn our attention to m. Is it odd or even? It cannot be even because n and m would then both be even and we would be able to simplify the fraction n/m by dividing both terms by 2. This, however, would contradict our original assumption that n/m had already been reduced to its lowest terms. We must assume, then, that m is odd.

Since n is even, we can express n in the form $2a$, letting a stand for another integer. Substituting $2a$ for n in equation (2), we have

$$4a^2 = 2m^2 , \qquad (3)$$

which reduces to

$$2a^2 = m^2 . \qquad (4)$$

By the same argument used above, m cannot be odd because its square equals the even number expressed by the left side of the equation. We previously saw that m cannot be even. Now we see that it also cannot be odd. Since every integer must be even or odd, m cannot be an integer. Our initial assumption must be false; there is no common fraction n/m that is the square root of 2. The number we seek is irrational, a term that reflects the shock of the discovery of such numbers by the ancient Greeks. Note also that equation (2) has been shown to be one that cannot be satisfied by integers. In other words, no square integer is exactly twice another square integer. This too is an important theorem that would be hard to prove without the astonishing power of a simple parity check.

In every branch of mathematics an odd-even check often supplies an efficient, short-cut proof. The following problem in topology is typical. Draw as many circles as you like, of any size, wherever you wish on a sheet of paper. Can such a "map" always be colored with two colors in such a way that no two regions with a common border are the same color? One way to prove that the answer is yes is to consider any pair of adjacent regions, A and B. They will be divided by an arc of a circle, which we will call X. One of the regions will be inside X, the other outside. Aside from X, A and B will be inside either no circles or the same number of circles; therefore one of the two regions is sure to be inside one more circle than the other. If we label each region with the number of circles it is within [*see Figure 57*], one of every pair of adjacent regions is sure to be even and its partner is sure to be odd. We color the even-numbered regions one color and the odd regions another color and the job is done. (For a different way of confirming the yes answer see my book *New Mathematical Diversions from Scientific American;* New York: Simon and Schuster, 1966; chapter 10.)

In the physical world things frequently have a mathematical pattern to which the familiar properties of odd and even numbers apply. An amusing instance is supplied by a parlor trick with three empty drinking

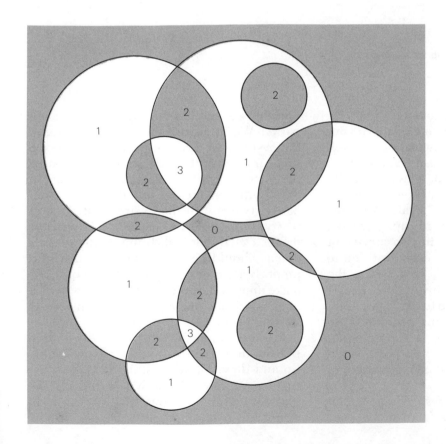

57. *A two-color map theorem*

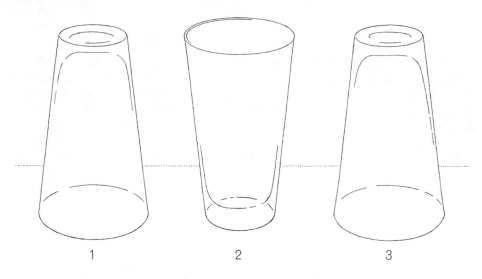

58. Setup for the glass trick

glasses. Place the glasses as shown in Figure 58. The puzzle, you explain to your audience, is to turn over two glasses simultaneously, one with each hand, and in three such "moves" bring all the glasses upright. To demonstrate: Turn glasses *1* and *2*, then *1* and *3*, then *1* and *2* again. All three glasses will then be right side up. (Actually, you can get them all up in two moves, or even one move, but you do it in three to confuse your spectators.) Now comes the sneaky part. Casually invert the center glass and invite someone to try. Few people will notice that the starting position is no longer the same as before. A simple parity check shows that from this new position the problem cannot be solved in *any* number of moves.

The proof is as follows. Whenever an even number of glasses (zero or two) are upright, we say that the "system" has even parity. When an odd number are upright, the system has odd parity. It is easy to see that turning any two glasses cannot change the system's parity. No amount of turning in pairs will convert the initial state of even parity (two up) to the desired state of odd parity (three up). If a spectator follows your moves exactly, he will bring all the glasses *down*. Should he accidentally set them up properly for a new attempt, step in quickly, make another fast demonstration of how it is done and leave him again with the incorrect starting position.

If there are ten glasses (or any even number not divisible by 4) arranged alternately up and down, is it possible to make a sequence of moves that will bring all the glasses up or all down? No, because in either case an impossible change in the system's parity (from, say, an odd five to an even ten) is demanded.

As long as the glasses behave politely, according to our notion of their structure, it is inconceivable that this parity-conservation law would be violated. But nature, particularly on the subatomic level, is under no obligation to conform to our notions of structure. In 1957 a parity law that for thirty years had been found applicable to the wave functions of quantum mechanics turned out not to hold in the case of the weak interactions of particles. Physicists are still recovering from the shock. It was as if someone had stepped up to ten alternating glasses, turned them in pairs and brought all ten upright!

An entertaining coin trick of the extra-sensory-perception variety exploits the same underlying principle as the glass trick. Someone is asked to take a handful of coins from his pocket and toss them on the table. While you look away, ask him to turn over the coins at random but always two coins simultaneously. He continues as long as he pleases, doing it silently so that you have no idea how many turns he makes. He then covers one coin with his hand. You turn around, glance at the other coins and immediately tell him whether the concealed coin is heads or tails.

The method (explained by Al Thatcher in the October, 1962, issue of a magician's periodical, *The New Phoenix*) could not be simpler. At the outset an even number of heads indicates even parity; an odd number, odd parity. If coins are turned in pairs, parity must be conserved. For example,

suppose five heads show at the beginning. At the finish, when one coin is hidden, the parity of the system must still be odd. Thus if you see an even number of heads, you know the concealed coin is a head. If you see an odd number of heads, the concealed coin must be a tail.

As a variation, let a spectator cover *two* coins and then tell him whether they match or not. Another variant is to let him turn first one coin, then two, then three, and repeat this triplet of changes as long as he wishes. Since $1 + 2 + 3 = 6$, an even number, parity will be conserved as before.

Sometimes the underlying parity structure of a system is so well camouflaged that only the most alert mathematician is able to spot it. A sterling example is provided by the following unusual problem adapted from *Unterhaltsame Mathematik,* a brilliant collection of puzzles by the German Mathematician Roland Sprague. (An English translation by T. H. O'Beirne was published in London in 1963 by Blackie and Son Ltd.) Five alphabet blocks, all exactly alike and each with the letter A on one face only, are first placed on a checkerboard in the cross formation shown in Figure 59, upper left corner of the board. The A sides of all five blocks are uppermost. The blocks are now moved from square to square by being tipped over along one edge, as one might move a large, heavy cubical box. In other words, each block is moved by a series of quarter turns each of which tips it from one square to an adjacent square. It is impossible, if one moves the blocks in this fashion, to arrange them in a row, anywhere

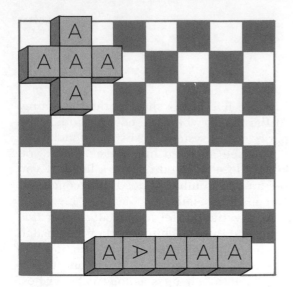

on the board, with the A faces uppermost and all with the same orientation. It *is* possible to arrange them as shown in the row at the bottom of the board. Which block in this row started out as the center block of the cross formation?

One could, of course, obtain five alphabet blocks and find by actual test which block it must be, but with the right insight into the odd-even structure of the system the correct block can be identified simply by studying the picture. Moreover, the parity check provides a proof the empirical test does not. The test merely shows that one block in the row *could have been* the center one; it does not prove that no other block could have been if the right sequence of turns had been made.

Perhaps an easier odd-even problem concerns the work of an eccentric U.S. architect, Frank Lloyd Wrong. To annoy a wealthy client, Wrong designed a house

shaped like an enormous shoe box. It was divided by floors into three levels, and on each level the floors were divided by vertical walls into seven rectangular rooms. There were no hallways, staircases, or closets, there was no basement or attic. The house consisted entirely of 21 rectangular, box-shaped rooms.

The doors of the house were of two types:

1. Conventional doors that enabled one to go from one room to a neighboring room, or from a first-floor room to the grounds outside.

2. Trap doors that allowed one, with the aid of ladders, to go from one room to a room directly above or below.

The doors were placed at random. One room might contain a dozen or more doors or (like the Other Professor's room in Lewis Carroll's *Sylvie and Bruno*) as few as no doors at all. Wrong was careful, however, to see that each room had an even number of doors. (Zero is considered even.) The problem is to prove that the number of outside doors, leading from first-floor rooms to the grounds, is even.

Answers

Which one of the five alphabet blocks in a row on a chessboard had been the center block in the previous formation before the blocks were moved by tipping them over an edge from square to square? It is obvious that if a block is moved an even number of times, it will rest on a square that is the same color as the square on which it started. An odd number of moves puts it on a square of opposite color. Not so obvious is the way in which odd and even apply to the orientations of each block.

Imagine a block painted red on three sides that meet at one corner and placed so you can see three of its sides. There are four possibilities: you see no red side, one red side, two red sides, or three red sides. If you see one or three red sides, we say the block has odd parity; otherwise, it has even parity. Whenever the block is given a quarter-turn in any direction, it is sure to change parity as shown in Figure 60. (This follows from the fact that opposite sides of the block are different colors. Each quarter-turn takes one side out of your line of vision and brings its opposite side into view. Thus a quarter-turn always alters one of the visible colors.)

Think of a block as a die instead of a block with colored sides. In this case its parity is indicated by whether the sum of the three visible faces is odd or even.

Because each move of the block gives it a quarter-turn, it changes its parity with each move. After an even number of moves it will be on a square of the same color as the square from which it started, and it will have the same parity. After an odd number of moves it will have changed both color of square and parity. The center block originally rested on white. If it moved an odd number of times, it will be in the second formation on a black square, its parity altered. But all the blocks on black squares in the second formation have the same parity, therefore the center block is not among them. It must have moved an *even* number of times. This would put it on a white square, with its parity the same as before. Of the two blocks on white squares, only the second from the right has unaltered

60. How a quarter-turn changes the parity of a cube

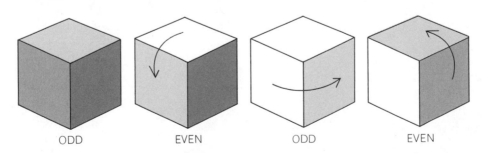

ODD   EVEN   ODD   EVEN

parity. Therefore it is the block we seek.

Actually, the blocks can be moved randomly to *any* final spots on the board and you can always identify the block that was originally at the center of the cross. It will be either the only block on a white cell with unaltered parity, or the only block on a black cell with altered parity.

To prove that Frank Lloyd Wrong's shoe-box house has an even number of outside doors, we consider first the fact that every door has two sides. If there are *n* doors, the total number of sides is $2n$, an even number. We are told that every room has an even number of doors. Assume that all doors are closed. An even number of *sides* will face into each room, therefore the total number of sides facing into rooms will be even. We subtract this even number from the total number of sides, also even, to obtain another even number: the number of sides

not facing into a room. These sides must, of course, be on the exterior doors. Therefore the number of doors leading to the grounds is even.

References

Informal Deduction in Algebra: Properties of Odd and Even Numbers. Commission on Mathematics of the College Entrance Examination Board, Princeton, N.J., 1959.

"Parity." Charles J. Goebel. In *McGraw-Hill Encyclopedia of Science and Technology.* Vol. 9. New York: McGraw-Hill, 1960. Pages 565–567.

Mathematics: The Man-made Universe. 2nd Edition. Sherman K. Stein. San Francisco: W. H. Freeman and Company, 1969. Chapters 1 and 2.

9. Patterns and Primes

No branch of number theory is more saturated with mystery and elegance than the study of prime numbers: those exasperating, unruly integers that refuse to be divided evenly by any integers except themselves and 1. Some problems concerning primes are so simple that a child can understand them and yet so deep and far from solved that many mathematicians now suspect they *have* no solution. Perhaps they are "undecidable." Perhaps number theory, like quantum mechanics, has its own uncertainty principle that makes it necessary, in certain areas, to abandon exactness for probabilistic formulation.

The central difficulty is that the primes are scattered along the series of integers in a pattern that clearly is not random and yet defies all attempts at precise description. What is the 100th prime? The only way a mathematician can answer is by obtaining a list of primes and counting to the 100th. How is such a list obtained initially? The simplest method is to go through the integers and cross out all the composite (not prime) numbers. Of course a computer can do this with great speed, but it still must use essentially the same simple-minded procedure that Eratosthenes, the Alexandrian geographer-astronomer and friend of Archimedes, devised two thousand years ago.

There is no better way to become familiar with the primes than by using Eratosthenes' Sieve (as his procedure is called) for sifting out all primes under 100. Kenneth P. Swallow of Monterey, California, has proposed an efficient way to do this. Write the numbers from 1 to 100 in the rectangular array shown in Figure 61. Cross out all multiples of 2, except 2 itself, by drawing vertical lines down the second, fourth and sixth columns. Eliminate the remaining multiples of 3 by drawing a line down the third column. The next integer not crossed out is 5. Multiples of 5 are removed by a series of diagonal lines running down and to the left. Remaining multiples of 7 are eliminated by lines sloping the other way. The integers 8, 9, and 10 are composite: their multiples have already been crossed out. Our job is now

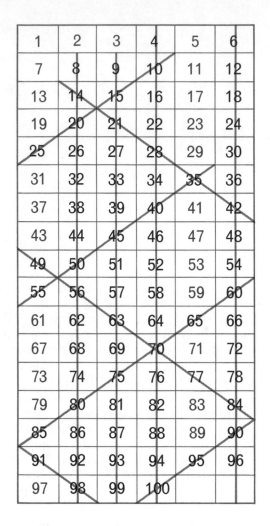

1	2	3	4	5	6
7	8	9	10	11	12
13	14	15	16	17	18
19	20	21	22	23	24
25	26	27	28	29	30
31	32	33	34	35	36
37	38	39	40	41	42
43	44	45	46	47	48
49	50	51	52	53	54
55	56	57	58	59	60
61	62	63	64	65	66
67	68	69	70	71	72
73	74	75	76	77	78
79	80	81	82	83	84
85	86	87	88	89	90
91	92	93	94	95	96
97	98	99	100		

61. The Sieve of Eratosthenes

finished because the next prime, 11, is larger than the square root of 100, the highest number in the table. Had the table been longer, larger multiples of 11 would have been removed by diagonal lines of steeper slope.

All but 26 numbers (shown in color) have fallen through the sieve. These are the first 26 primes. Mathematicians prefer to say 25 primes, because various important theorems are simpler to express if 1 is not called a prime. For example, the "fundamental theorem of arithmetic" states that every integer greater than 1 can be factored into a unique set of prime numbers. Thus 100 is the product of four primes: $2 \times 2 \times 5 \times 5$. No other set of positive primes has a product of 100. If 1 were called a prime, we could not say this. There would be an infinite number of different sets of prime factors, such as $2 \times 2 \times 5 \times 5 \times 1 \times 1$.

Much can be learned about the primes by studying Figure 61. You see at once that all primes greater than 3 are either one less or one more than a multiple of 6. Also, it is clear why there are so many "twin primes": pairs of primes that have a difference of 2, such as 71 and 73, 209,267 and 209,269, or 1,000,000,009,649 and 1,000,000,009,651. After eliminating multiples of 2 and 3, *all* remaining numbers are twin-paired. Subsequent sievings simply remove one or both partners of a pair, but they leave many untouched. Twin primes get scarcer as the numbers get bigger. It is conjectured that an infinity of them continue to sift through the sieve, but no one knows for certain. The chart also shows at a glance that 3, 5, 7 is the only possible triplet of primes.

If the integers are differently placed, the primes will of course form a different geometrical pattern. In 1963 Stanislaw M. Ulam, of the Los Alamos Scientific Laboratory, attended a scientific meeting at which he found himself listening to what he describes as a "long and very boring paper." To pass the time he doodled a grid of horizontal and vertical lines on a sheet of paper. His first impulse was to compose some chess problems, then he changed his mind and began to number the intersections, starting near the center with 1 and moving out in a counterclockwise spiral. With no special end in view, he began circling all the prime numbers. To his surprise the primes seemed to have an uncanny tendency to crowd into straight lines. Figure 62 shows how the

primes appeared on the spiral grid from 1 to 100. (For clarity the numbers are shown inside cells instead of on intersections.)

Near the center of the spiral the lining up of primes is to be expected because of the great "density" of primes and the fact that all primes except 2 are odd. Number the squares of a checkerboard in spiral fashion and you will discover that all odd-numbered squares are the same color. If you take 17 checkers (to represent the 17 odd primes under 64) and place them at random on the

62. *Ulam's square spiral*

100	99	98	97	96	95	94	93	92	91
65	64	63	62	61	60	59	58	57	90
66	37	36	35	34	33	32	31	56	89
67	38	17	16	15	14	13	30	55	88
68	39	18	5	4	3	12	29	54	87
69	40	19	6	1	2	11	28	53	86
70	41	20	7	8	9	10	27	52	85
71	42	21	22	23	24	25	26	51	84
72	43	44	45	46	47	48	49	50	83
73	74	75	76	77	78	79	80	81	82

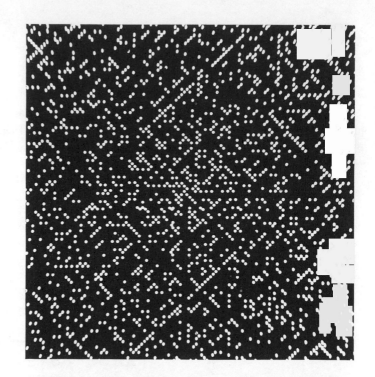

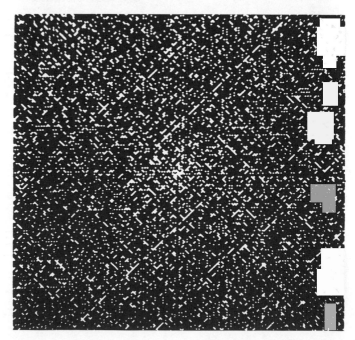

63. Photographs of a computer
grid showing primes as
a spiral of integers from
1 to about 10,000 (top)
and from 1 to about 65,000
(bottom)

32 odd-numbered squares, you will find that they form diagonal lines. But in the higher, less dense areas of the number series one would not expect many such lines to form. How would the grid look, Ulam wondered, if it was extended to thousands of primes?

The computer divison at Los Alamos has a magnetic tape on which 90 million prime numbers are recorded. Ulam, together with Myron L. Stein and Mark B. Wells, programed the MANIAC computer to display the primes on a spiral of consecutive integers from 1 to about 65,000. The picture of the grid presented by the computer is shown in Figure 63. Note that even near the picture's outer limits the primes continue to fall obediently into line.

The eye first sees the diagonally compact lines, where odd-number cells are adjacent, but there is also a marked tendency for primes to crowd into vertical and horizontal lines on which the odd numbers mark every other cell. Straight lines in all directions (once they have been extended beyond the consecutive numbers on a segment of the spiral) bear numbers that are the values of quadratic expressions beginning with $4x^2$. For example, the diagonal sequence of primes 5, 19, 41, 71 is given by the expression $4x^2 + 10x + 5$ as x takes the values 0 through 3. The grid suggests that throughout the entire number series expressions of this form are likely to vary markedly from those "poor" in primes to those that are "rich," and that on the rich lines an unusual amount of clumping occurs.

By starting the spiral with numbers higher than 1 other quadratic expressions form the lines. Consider a grid formed by starting the spiral with 17 [*see Figure 64, left*]. Numbers in the main diagonal running northeast by southwest are generated by $4x^2 + 2x + 17$. Plugging positive integers into x gives the diagonal's lower half; plugging negative integers give the upper half. If we consider the entire diagonal, rearranging the numbers in order of increasing size, we find — pleasantly enough — that all the numbers are generated by the simpler formula $x^2 + x + 17$. This is one of many "prime-rich" formulas discovered by Leonhard Euler, the eighteenth-century Swiss mathematician. It generates primes for all values of x from 0 through 15. This means that if we continue the spiral shown in the illustration until it fills a 16-by-16 square, the entire diagonal will be solid with primes.

Euler's most famous prime generator, $x^2 + x + 41$, can be diagramed similarly on a spiral grid that starts with 41 [*see Figure 64, right*]. This produces an unbroken sequence of 40 primes, filling the diagonal of a 40-by-40 square! It has long been known that of the first 2,398 numbers generated by this formula, exactly half are prime. After testing all such numbers below 10,000,000, Ulam, Stein, and Wells found the proportion of primes to be .475 . . . Mathematicians would like to have a formula expressing a function of n that would give a different prime for *every* integral value of n. It has been proved that no polynomial formula of this type exists. There are many nonpolynomial formulas that *will* generate only

33	32	31	30	29
34	21	20	19	28
35	22	17	18	27
36	23	24	25	26
37	38	39		

57	56	55	54	53
58	45	44	43	52
59	46	41	42	51
60	47	48	49	50
61	62	63		

*64. Diagonals generated by the formula $x^2 + x + 17$ (left)
and $x^2 + x + 41$ (right)*

primes, but they are of such a nature that they are of no use in computing primes because the sequence of primes must be known in order to operate with the formulas. (See "History of a Formula for Primes," by Underwood Dudley, *The American Mathematical Monthly*, January, 1969.)

Ulam's spiral grids have added a touch of fantasy to speculations about the enigmatic blend of order and haphazardry in the distribution of primes. Are there grid lines that contain an infinity of primes? What is the maximum prime density of a line? On infinite grids are there density variations between top and bottom halves, left and right, the four quarters? Ulam's doodlings in the twilight zone of mathematics are not to be taken lightly. It was he who made the suggestion that led him and Edward Teller to think of the "idea" that made possible the first thermonuclear bomb.

Although primes grow steadily rarer as numbers increase, there is no highest prime.

The infinity of primes was concisely and beautifully proved by Euclid. One is tempted to think, because of the rigidly ordered procedure of the sieve, that it would be easy to find a formula for the exact number of primes within any given interval on the number scale. No such formula is known. Early nineteenth-century mathematicians made an empirical guess that the number of primes under a certain number *n* is approximately *n*/natural log of *n*, and that the approximation approaches a limit of exactness as *n* approaches infinity. This astonishing theorem, known as the "prime-number theorem," was rigorously proved in 1896. (See "Mathematical Sieves," by David Hawkins, *Scientific American*, December, 1958, for a discussion of this theorem and its application to other types of numbers, including the "lucky numbers" invented by Ulam.)

It is not easy to find the mammoth primes isolated in the vast deserts of composite numbers that blanket ever larger areas of the number series. At the moment the largest known prime is $2^{19937} - 1$, a number of 6,002 digits. It was discovered in 1971 by Bryant Tuckerman, at IBM's research center, Yorktown Heights, New York. Before the advent of modern computers, testing a number of only six or seven digits could take weeks of dreary calculation. Euler once announced that 1,000,009 was prime, but he later discovered that it is the product of two primes: 293 and 3,413. This was a considerable feat at the time, considering that Euler was 70 and blind. Pierre Fermat was once asked in a letter if 100,895,598,169 is prime. He shot back that it is the product of primes 898,423 and 112,303. Feats such as these have led some to think that the old masters may have had secret and now-lost methods of factoring. As late as 1874 W. Stanley Jevons could ask, in his *Principles of Science:* "Can the reader say what two numbers multiplied together will produce the number 8,616,460,799? I think it unlikely that anyone but myself will ever know; for they are two large prime numbers." Jevons, who himself invented a mechanical logic machine, should have known better than to imply a limit on future computer speeds. Today a computer can find his two primes (96,079 and 89,681) faster than he could multiply them together.

Numbers of the form $2^p - 1$, where p is prime, are called Mersenne numbers after Marin Mersenne, a seventeenth-century Parisian friar (he belonged to a humble order known as the Minims — an appropriate order for a mathematician), who was the first to point out that many numbers of this type are prime. For some 200 years the Mersenne number $2^{67} - 1$ was suspected of being prime. Eric Temple Bell, in his book *Mathematics, Queen and Servant of Science*, recalls a meeting in New York of the American Mathematical Society in October, 1903, at which Frank Nelson Cole, a Columbia University professor, rose to give a paper. "Cole — who was always a man of very few words — walked to the board and, saying nothing, proceeded to chalk up the arithmetic for raising 2 to the sixty-seventh power. Then he carefully subtracted 1. Without a word he moved over to a clear space on the board and multiplied out, by longhand,

$$193,707,721 \times 761,838,257,287.$$

The two calculations agreed. . . . For the first and only time on record, an audience of the American Mathematical Society vigorously applauded the author of a paper delivered before it. Cole took his seat without having uttered a word. Nobody asked him a question." Years later, when Bell asked Cole how long it took him to crack the number, he replied, "Three years of Sundays."

The British puzzle expert Henry Ernest Dudeney, in his first puzzle book (*The Canterbury Puzzles*, 1907), pointed out that 11 was the only known prime consisting entirely of 1's. (Of course, a number formed by repeating any other digit would be composite.) He was able to show that all such "repunit" numbers, from 3 through 18 units, are composite. Are any larger "repunit" chains prime? Oscar Hoppe, a New York City reader of Dudeney's book, took up the challenge and actually managed to prove, in 1918, that the 19-"repunit" number, 1,111,111,111,111,111,111 is prime. Later it was discovered that 23 repeated 1's is also prime. There the matter rests. No one knows if the "repunit" primes are infinite, or even if there are more than three. It is easy to see that no repunit number is prime unless the number of its units is prime. (For example, if its number of digits has a factor of, say, 13, then clearly it is divisible by a repunit of 13 digits.) As of 1970 repunits have been tested through 373 digits without finding a fourth prime.

Can a magic square be constructed solely of different primes? Yes; Dudeney was the first to do it. Figure 65 shows such a square. It sums in all directions to the "repunit" number 111: the lowest possible constant for a prime square of order 3. (Curiously, an order-4 square is possible with the lower magic constant of 102. See Dudeney's *Amusements in Mathematics;* New York: Dover, 1917; problem 408.)

Can a magic square be made with *consecutive* odd primes? (The even prime, 2, must be left out because it would make the odd or even parity of its rows and columns different from the parity of all other rows and columns, thereby preventing the array from being magic.) In 1913 J. N. Muncey of Jessup, Iowa, proved that the smallest magic square of this type is one of order 12. This remarkable curiosity is so little known that I reproduce it in Figure 66. Its cells

65. Prime magic square with lowest order-3 constant

hold the first 144 consecutive odd primes, starting with 1. All rows, columns and the two main diagonals sum to 4,514.

Readers may test their familiarity with primes by answering the following elementary questions:

1. Identify the four primes among the following six numbers. (*Note:* The second number is the first five digits in the decimal of pi.)

10,001

14,159

76,543

77,377

123,456,789

909,090,909,090,909,090,-909,090,909,091

1	823	821	809	811	797	19	29	313	31	23	37
89	83	211	79	641	631	619	709	617	53	43	739
97	227	103	107	193	557	719	727	607	139	757	281
223	653	499	197	109	113	563	479	173	761	587	157
367	379	521	383	241	467	257	263	269	167	601	599
349	359	353	647	389	331	317	311	409	307	293	449
503	523	233	337	547	397	421	17	401	271	431	433
229	491	373	487	461	251	443	463	137	439	457	283
509	199	73	541	347	191	181	569	577	571	163	593
661	101	643	239	691	701	127	131	179	613	277	151
659	673	677	683	71	67	61	47	59	743	733	41
827	3	7	5	13	11	787	769	773	419	149	751

66. *Smallest possible magic square of consecutive odd primes*

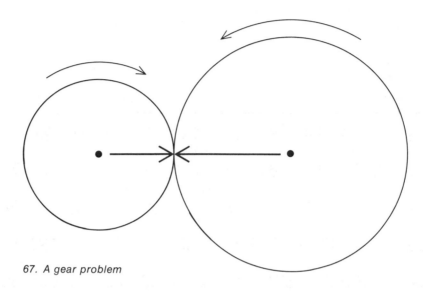

67. A gear problem

2. Two gear wheels, each marked with an arrow, mesh as shown in Figure 67. The small wheel turns clockwise until the arrows point directly toward each other once more. If the large wheel has 181 teeth, how many times will the small wheel have rotated? (Contributed by Burris Smith of Greenville, Mississippi.)

3. Using each of the nine digits once, and only once, form a set of three primes that have the lowest possible sum. For example, the set 941,827, and 653 sum to 2,421, but this is far from minimal.

4. Find the one composite number in the following set:

31 331 3331 33331 333331 3333331
33333331 333333331

5. Find a sequence of a million consecutive integers that contains not a single prime.

Addendum

Many *Scientific American* readers experimented with triangular and hexangular arrays of integers and found that the primes cluster along straight lines in the same manner as in Stanislaw Ulam's square spiral grids. Laurence M. Klauber of San Diego, California, sent me a copy of a paper he had read to a meeting of the Mathematical Association of America in 1932, discussing his search for prime-rich polynomial formulas in such an array. Ulam has also used the Los Alamos computer for investigating a variety of other types of grid, including the triangular, and in every case he found that significant departures from random distributions of primes were at once apparent. This is hardly surprising, because any orderly arrangement of consecutive integers in a grid will have straight lines that are generated by polynomial expressions. If the expression is factorable, the line cannot contain primes; this fact alone can account for a concentration of primes along certain other lines.

All diagonals of even numbers are obviously prime-empty, and other lines are empty because they are factorable by other numbers. Many readers noticed that the diagonal line extending down and to the right from 1 on Ulam's spiral grid contains in sequence the squares of odd integers, and the diagonal line extending up and to the left from 4 gives the squares of even integers. Both diagonals are, of course, prime-empty. Conversely, other lines are prime-rich because they are generated by formulas that act as sieves, removing numbers that are multiples of low primes. The significance of Ulam's spiral grids lies not in the discovery that primes are nonrandomly distributed, which is to be expected in any orderly arrangement of integers, but in the use of a computer and scope to extend such grids quickly so that photographs provide, so to speak, a bird's-eye view of the pattern from which hints can be obtained that may lead to new theorems.

Several readers called my attention to W. H. Mills's formula in the *Bulletin of the American Mathematical Society*; June, 1947, page 604, which contains an irrational number between 1 and 2. When positive integers are substituted for n in the formula, the expression gives prime values; but since the irrational number is not known, the formula is of no value in computing primes. In fact, it is easy to write irrational numbers that generate every prime in sequence, for example .2030507011013017-0190230. . . . To be sure, one has to know the sequence of primes before computing the number. There are many ways of writing complicated functions of n so that integral values of n produce distinct primes, but the catch is that the function itself requires the introduction of the prime-number sequence, making the formula valueless for finding primes. Readers interested in formulas of this type will find a nontechnical discussion of them in Oystein Ore's excellent book *Number Theory and Its History* (New York: McGraw-Hill, 1948).

Answers

1. The two composite numbers are 10,001 (the product of primes 73 and 137) and 123,456,789, which is evenly divisible by 3. The other numbers are primes.

2. Two meshed gear wheels of different sizes cannot return to the same position until a certain number of teeth, k, have passed the point of contact on both wheels. The number k is the lowest common multiple of the number of teeth on each wheel. Let n be the number of teeth on the small wheel. We are told that the large wheel has 181 teeth. Since 181 is a prime number, the lowest common multiple of n and 181 is $181n$. Therefore the small wheel will have to make 181 rotations before the two wheels will return to their former position.

3. How can the nine digits be arranged to make three primes with the lowest pos-

sible sum? We first try numbers of three digits each. The end digits must be 1, 3, 7, or 9 (this is true of all primes greater than 5). We choose the last three, freeing 1 for a first digit. The lowest possible first digits of each number are 1, 2, and 4, which leaves 5, 6, and 8 for the middle digits. Among the 11 three-digit primes that fit these specifications it is not possible to find three that do not duplicate a digit. We turn next to first digits of 1, 2, and 5. This yields the unique answer

$$
\begin{array}{r}
149 \\
263 \\
587 \\
\hline
999
\end{array}
$$

4. The last number, 333333331, has a factor of 17. (The problem is based on a result obtained by Andrzej Makowski of Poland, which was reported in *Recreational Mathematics Magazine* for February, 1962.)

5. It is easy to find as large an interval as we please of consecutive integers that are not prime. For an interval of a million integers, consider first the number 1,000,001! The exclamation mark means that the number is "factorial 1,000,001," or the product of $1 \times 2 \times 3 \times 4 \ldots \times 1,000,001$. The first number of the interval we seek is 1,000,001! + 2. We know that 1,000,001! is divisible by 2 (one of its factors), so that if we add another 2 to it, the resulting integer must also be divisible by 2. The second number of the interval is 1,000,001! + 3. Again, because 1,000,001! has a factor of 3, it must be divis-

ible by 3 after we add 3 to it. Similarly for 1,000,001! + 4, and so on up to 1,000,001! + 1,000,001. This gives a consecutive sequence of one million composite numbers.

Are these the smallest integers that form a sequence of one million nonprimes? No, as Ted L. Powell pointed out in *The Graham Dial* for April, 1960; we can obtain a lower sequence just as easily by *subtracting:* 1,000,001! − 2; 1,000,001! − 3; and so on to 1,000,001! − 1,000,001.

References

"Magic Squares Made with Prime Numbers to Have the Lowest Possible Summations." W. S. Andrews and Harry A. Sayles. *The Monist*, Vol. 23, No. 4; October, 1913. Pages 623–630.

History of the Theory of Numbers: Volume I. Leonard Eugene Dickson. Carnegie Institution, 1919. (Reprint. Bronx, N.Y.: Chelsea Publishing Co., 1952.)

"The Factorgram." Kenneth P. Swallow. *The Mathematics Teacher*, Vol. 48, No. 1; January, 1955. Pages 13–17.

The First Six Million Prime Numbers. C. L. Baker and F. J. Gruenberger. Madison, Wis.: Microcard Foundation, 1959.

"A Visual Display of Some Properties of the Distribution of Primes." M. L. Stein, S. M. Ulam, and M. B. Wells. *The American Mathematical Monthly*, Vol. 71, No. 5; May, 1964. Pages 516–520.

"Peculiar Properties of Repunits." Samuel Yates. *Journal of Recreational Mathematics*, Vol. 2, No. 3; July, 1969. Pages 139–146.

10. Graph Theory

AN ENGINEER draws a diagram of an electrical network. A chemist makes a sketch to show how the atoms of a complex molecule are joined by chemical bonds. A genealogist draws an intricate family tree. A military commander plots a network of supply lines on a map. A sociologist traces in an elaborate diagram the power structure of a giant corporation.

What do all these patterns have in common? They are points (representing electrical connections, atoms, people, cities, and so on) connected by lines. In the 1930's the mathematician Dénes König made the first systematic study of all such patterns, giving them the generic name "graphs." (The confusion of this term with the "graphs" of analytic geometry is regrettable, but the term has stuck.) Today graph theory is a flourishing field. It is usually considered a branch of topology (because in most cases only the topological properties of graphs are considered), although it now overlaps large areas of set theory, combinatorial mathematics, algebra, geom-

etry, matrix theory, game theory, logic, and many other fields.

König's pioneer book on graphs (published in Leipzig in 1936) has yet to be translated, but an English edition of a later French book, *The Theory of Graphs and Its Applications*, by Claude Berge, was published in England in 1962. Oystein Ore's elementary introduction, *Graphs and Their Uses*, was issued as a paperback (New York: Random House, 1963). Both books are of great recreational interest. Hundreds of familiar puzzles, seemingly unrelated, yield readily to graph theory. In this chapter we center our attention on "planar graphs" and some of their more intriguing puzzle aspects.

A planar graph is a set of points, called vertices, connected by lines, called edges, in such a way that it is possible to draw the graph on a plane without any pair of edges intersecting. Imagine that the edges are elastic strings that can be bent, stretched, or shortened as we please. Is the graph shown in Figure 68 planar? (Its four ver-

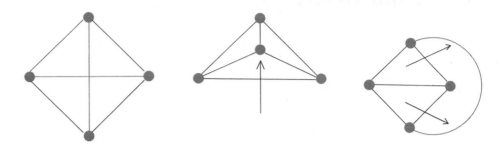

68. *Three ways to draw a complete graph for four points*

tices are indicated by spots. The crossing point at the center is not a vertex; think of one line as passing under the other.) Yes, because we can easily remove the intersection by shifting the position of a vertex, as shown in the middle graph, or stretching an edge as shown in the one at the right. All three of these diagrams are "isomorphic": each represents a different way of drawing the *same* planar graph. The edges of any solid polyhedron, such as a cube, are planar graphs because we can always stretch the solid's "skeleton" until it lies on a plane, free of intersections. The skeleton of a tetrahedron is isomorphic with the three graphs of Figure 68.

It is not always easy to decide if a graph is planar. Consider the problem depicted at the left in Figure 69, one of the oldest and most frustrating of all topological teasers. Since the English puzzlist Henry Ernest Dudeney gave it this form in 1917 it has been known as the "utilities problem." Each house must receive gas, water and electricity. Can lines be drawn to connect each house with each utility in such a way that no line intersects another? In other

words, is the resulting graph planar?

The answer is no, and it is not difficult to give a rough proof. Assume that only houses *A* and *B* are to be connected to the three utilities. To do this without having any line cross another you must divide the plane into three regions as shown at the right in Figure 69. Your lines need not be as pictured, but however you draw them your graph will be isomorphic with the one shown. House *C* must go in one of the three regions. If it goes in *X*, it is cut off from electricity. If it goes in *Y*, it is cut off from water. At *Z*, it is cut off from gas. The same argument holds if the graph is drawn on a sphere, but not if it is drawn on certain other surfaces. For example, the graph is easily drawn without intersections on the surface of a doughnut.

When every vertex of a graph is connected to each of the other vertices, the graph is said to be "complete." We saw in Figure 68 that the complete graph for four points is planar. Is the complete graph for five points planar? Again an informal proof (the reader may enjoy working it out for himself) shows it is not. This proof is equivalent to a proof

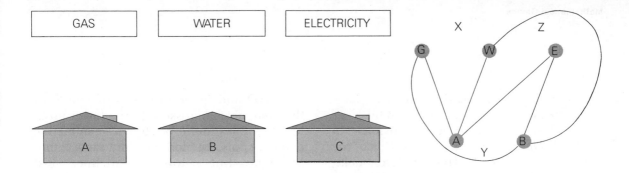

69. Problem of the three utilities (left) *and the impossibility proof for the utilities problem* (right)

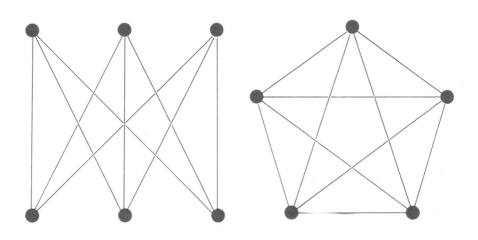

70. Simplest nonplanar graphs

that it is not possible to draw five regions in such a way that every pair shares a common border segment, a theorem often confused with the famous four-color map theorem. The two simplest nonplanar graphs are shown in Figure 70. At the left is the utilities graph (known as a Thomsen graph), at the right is the complete graph for five points.

The fact that a complete graph can be planar only if it has four or fewer points is not without philosophical interest. Many philosophers and mathematicians have tried to answer the question: Why does physical space have three dimensions? In his book *The Structure and Evolution of the Universe* (New York: Harper Torchbooks, 1959) the British cosmologist G. J. Whitrow argues that intelligent life as we know it could not have evolved in a space of *more* than three dimensions because such spaces do not allow stable planetary orbits around a sun. How about spaces of one or two dimensions? Intelligent Linelanders and Flatlanders are ruled out, says Whitrow, by graph theory. A brain requires an immense number of nerve cells (points), connected in pairs by nerves (edges) that must not intersect. In three dimensions there is no limit to the number of cells that can be so connected, but in a Flatland the maximum number, as we have seen, would be four.

"Thus," Whitrow writes, "we may conclude that the number of dimensions of physical space is necessarily three, no more and no less, because it is the unique natural concomitant of the evolution of the higher forms of terrestrial life, in particular of Man, *the formulator of the problem.*"

Devising planar graphs is an essential task in many fields of technology. Printed circuits, for instance, will short-circuit if any two paths cross. The reader may wish to test his skill in planar graph construction by considering the two printed-circuit problems shown in Figure 71. In the upper problem five nonintersecting lines must be drawn within the rectangle, each connecting a pair of spots bearing the same letter (*A* with *A*, *B* with *B*, and so on). The two

lines *AD* and *BC* are barriers of some sort that may not be crossed. In the lower problem five lines are to be drawn—connecting pairs of spots, labeled with the same letter, as before—but in this case all lines must follow the grid. Of course there must be no crossings. The solution is unique.

Another well-known type of graph puzzle is the one that calls for drawing a given planar graph in one continuous line without taking a pencil from the paper or going over any edge twice. If such a line can be drawn as a closed loop, returning to the vertex from which it started, the graph is said to be an "Euler graph" and the line an "Euler line." In 1736 the Swiss mathematician Leonhard Euler solved a famous problem involving a set of seven bridges in the East Prussian town of Königsberg (now Kaliningrad). Was it possible to walk over each bridge once and only once and return to where one had started? Euler found that the problem was identical with that of tracing a simple graph. He showed, in the first paper ever written on graph theory, that if every vertex of a graph is of "even degree" (has an even number of lines meeting it), it can be traced in one round-trip path. If there are two vertices of odd degree, no round trip is possible, but the graph can be drawn by a line beginning at one odd vertex and ending at the other. If there are $2k$ vertices of odd degree (and the number of odd vertices must always be even), it can be traced by k separate paths, each starting and ending at an odd vertex. The graph for the bridges of Königsberg has four odd vertices, therefore it requires a minimum of two paths (neither of

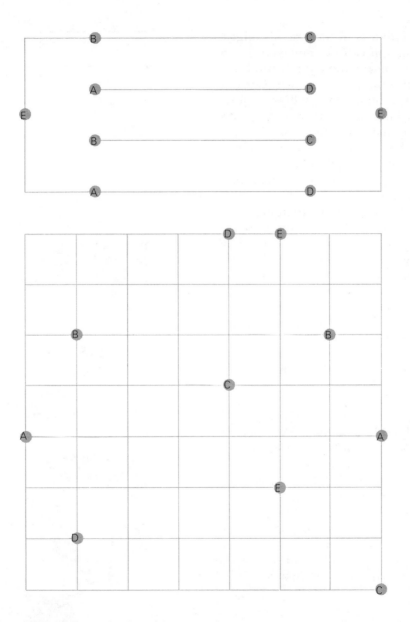

71. Two printed-circuit problems

them closed circuits) to traverse all edges.

Any Euler graph can be traversed by an Euler line that makes the entire round trip without intersecting itself. Lewis Carroll, we are told in a biography by his nephew, was fond of asking little girls to draw, with one Euler line, the graph in Figure 72. It is easily done if lines are allowed to intersect, but it is not so easy if intersections are forbidden. A quick way to solve such puzzles has been proposed by T. H. O'Beirne of Glasgow. One colors alternate regions as shown in the middle drawing, then breaks them apart at certain vertices in any way that will leave the colored areas "simply connected" (connected without enclosing noncolored areas). The perimeter of the colored region is now the Euler line we seek [*at bottom, right*]. The reader can try this method on the Euler graph shown in Figure 73 (proposed by O'Beirne) to see how pleasingly symmetrical an Euler line he can obtain.

An entirely different and, strangely, much more difficult type of graph-traversing puzzle is that of finding a route that passes through each vertex once and only once. Any route that passes through no vertex twice is known in graph theory as an arc. An arc that returns to the starting point is called a circuit. And a circuit that visits every vertex once and only once is called a Hamiltonian line, after Sir William Rowan Hamilton, the nineteenth-century Irish mathematician, who was the first to study

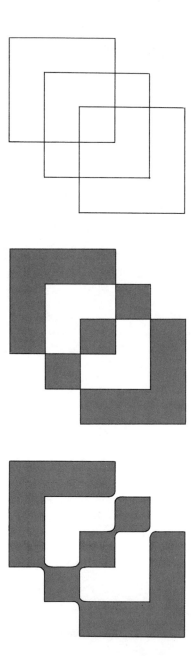

72. Lewis Carroll's three-square problem

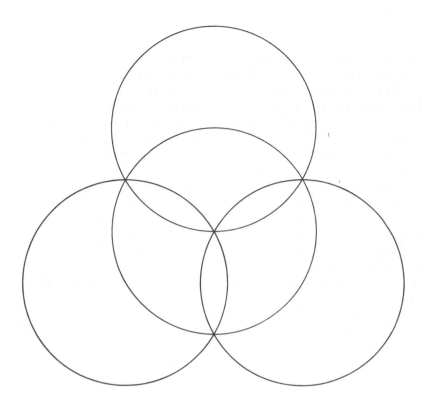

73. O'Beirne's four-circle problem

such paths. He showed that a Hamiltonian line could be traced along the edges of each of the five regular solids, and he even sold a toy manufacturer a puzzle based on finding Hamiltonian tours along the edges of the dodecahedron.

It might be supposed that, as in the case of Euler lines, there would be simple rules for determining if a graph is Hamiltonian; the fact is that the two tasks are surprisingly dissimilar. An Euler line must trace every edge once and only once, but it may go through any vertex more than once. A Hamiltonian line must go through each vertex once and only once, but it need not trace every edge. (In fact, it traverses exactly two of the edges that meet at any one vertex.) Hamiltonian paths are important in many fields where one would not expect to find them. In operations research, for example, the problem of obtaining the best order in which to carry out a specified series of operations can sometimes be diagramed as a graph on which a Hamiltonian line gives an optimum solution. Unfortunately there is no general method for deciding if a graph is Hamiltonian, or for finding all Hamiltonian lines if it is.

Many semiregular polyhedrons, but not all, have Hamiltonian skeletons. An exception is the rhombic dodecahedron shown in Figure 74, a form often assumed by crystals of garnet. Even if the path is not required to be closed, there is no way to traverse the skeleton so that each vertex is visited once and only once. The proof, first given by H. S. M. Coxeter, is a clever one. All vertices of degree 4 are shown as black spots, all of degree 3 as colored spots. Note that every black spot is completely surrounded by colored spots and vice versa. Therefore any path through all 14 spots must alternate

colored and black. But there are six black spots and eight colored ones. No path of alternating color is possible, either closed or open at the ends.

An ancient chess recreation that at first seems far removed from Hamiltonian paths is the reentrant knight's tour. It consists of placing the knight on a square of the chessboard, then finding a path of continuous knight's moves that will visit every square once and only once, the knight thereupon returning in one move to the square from which it started. Suppose each cell of the board to be represented by a point and every possible knight's move by a line joining two points. The result is, of course, a graph. Any circuit that visits each vertex once and only once will be a Hamiltonian line, and every such line will trace a reentrant knight's tour.

Such a tour is impossible on any board with an odd number of cells. (Can the reader see why?) The smallest rectangle on which a closed tour is possible is one with an area of 30 square units (3 × 10, or 5 × 6). The six-by-six is the smallest square. No tours, not even open-ended ones, are possible on rectangles with one side less than three. No one knows how many millions of different reentrant knight's tours can be made on the standard eight-by-eight chessboard. In the enormous literature on the topic the search has usually been confined to paths that exhibit interesting symmetries. Thousands of elegant patterns, such as those shown in Figure 75 have been discovered. Paths with exact fourfold symmetry (unchanged by any 90-degree rotation) are not

74. The skeleton of a rhombic dodecahedron

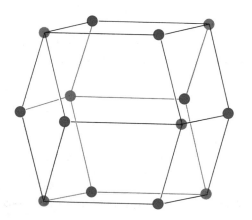

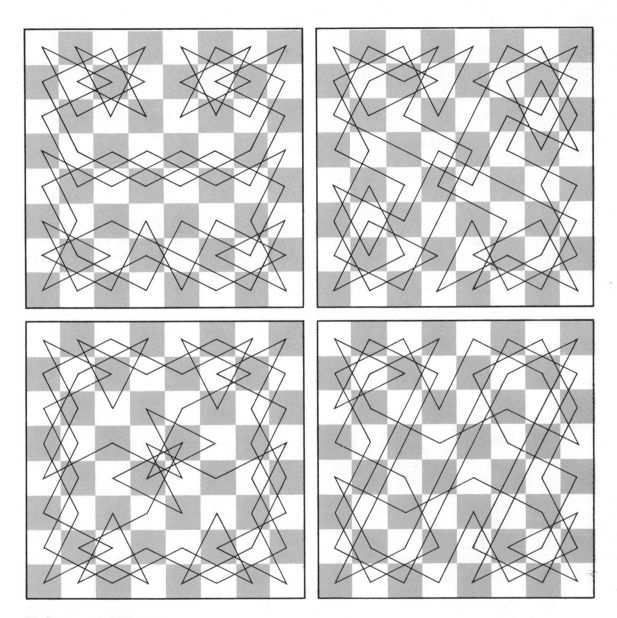

75. Reentrant knight's tours

possible on the eight-by-eight board, although five such patterns are possible on the six-by-six.

As an introduction to this classic pastime you are invited to search for a reentrant knight's tour on a simple 12-cell board [*see Figure 76*]. After it has been found, a seemingly more difficult question arises: Is it possible to move the knight over this board in one chain of jumps and make every possible knight's move once and only once? There are 16 different knight's moves. A move is considered "made" whenever a knight connects the two cells by a jump in either direction. Of course, the knight may visit any cell more than once, but it must not make the same move twice. The path need not be reentrant.

76. A knight's-tour problem

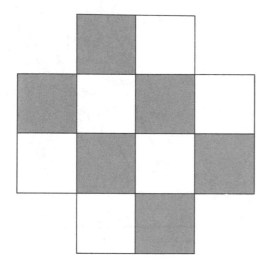

The reader will soon convince himself that such a path is not possible; but what is the smallest number of *separate* paths that will cover all 16 of the possible moves? This can be answered in just a few minutes by applying one of the graph theorems discussed earlier in this chapter.

Answers

The two printed-circuit problems are solved in the manner shown in Figure 77. A symmetrical, non-self-intersecting Euler line for the four-circle puzzle is shown in Figure 78, obtained by the coloring method explained on page 96. The path at the left in Figure 79 traces a reentrant knight's tour on the cross-shaped board. To determine if there is a single path that will go over every possible knight's move, we first draw a graph [*at right in illustration*] showing every move. Note that eight of the vertices are meeting points for an odd number of edges. In accordance with one of Euler's theorems, a minimum of 8/2, or 4, paths are required to trace every edge once and only once. Each path must begin at one odd vertex and end at another.

To prove that no reentrant knight's tour is possible on a board with an odd number of cells, first color the cells alternately, checkerboard fashion. Every knight's move carries the piece from a cell of one color to a cell of another, so that if the path is a closed circuit, half the cells in the path must be one color and half another color.

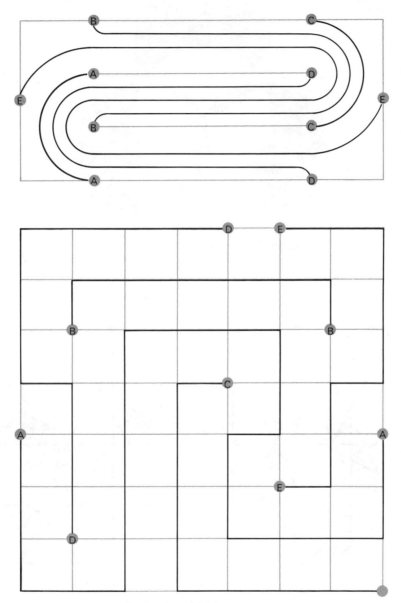

77. *Solutions to printed-circuit problems*

78. Solution to four-circle problem

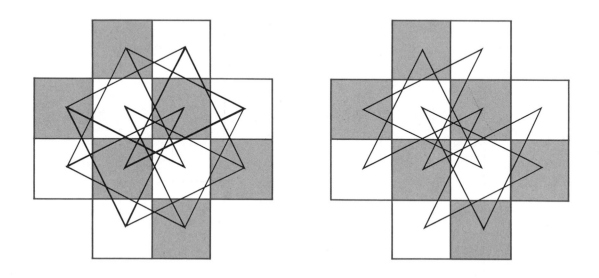

79. Graphs for reentrant knight's tour (left) and for all knight's moves (right)

But if a board has an odd number of cells, regardless of its shape there will be more cells of one color than of the other.

References

On the Traversing of Geometrical Figures. J. C. Wilson. Oxford: Oxford University Press, 1905.

Economic Applications of the Theory of Graphs. Giuseppe Avondo-Bodino. New York: Gordon and Breach, 1962.

The Theory of Graphs and Its Applications. Claude Berge. New York: Barnes and Noble, 1962.

Flows in Networks. L. R. Ford and D. R. Fulkerson. Princeton, N.J.: Princeton University Press, 1962.

Graphs and Their Uses. Oystein Ore. New York: Random House, 1963.

Groups and Their Graphs. Israel Grossman and Wilhelm Magnus. New York: Random House, 1964.

Finite Graphs and Networks: An Introduction with Applications. Robert G. Busaker and Thomas L. Sarty. New York: McGraw-Hill, 1965.

Structural Models: An Introduction to the Theory of Directed Graphs. Frank Harary, Robert Z. Norman, and Dorwin Cartwright. New York: John Wiley and Sons, 1965.

Connectivity in Graphs. W. H. Tutte. Toronto: University of Toronto Press, 1967.

A Seminar on Graph Theory. Edited by Frank Harary. New York: Holt, Rinehart and Winston, 1967.

11. The Ternary System

Somewhere in the darkness a woman sang in a high wild voice and the tune had no start and no finish and was made up of only three notes which went on and on and on.

Carson McCullers,
The Ballad of the Sad Café

NOW AND THEN a cultural anthropologist, eager to push mathematics into the folkways, will point to the use of different number systems in primitive societies as evidence that laws of arithmetic vary from culture to culture. But of course the same old arithmetic is behind every number system. The systems are nothing more than different languages: different ways of uttering, symbolizing, and manipulating the *same* numbers. Two plus two is invariably four in any notation, and it is always possible to translate perfectly from one number language to another.

Any integer except 0 can furnish the base, or radix, of a number system. The simplest notation, based on 1, has only one symbol: the notches an outlaw cuts in his gun or the beads a billiard player slides along a wire to record his score. The binary system has two symbols: 0 and 1. The decimal system, now universal throughout the civilized world, uses ten symbols. The larger the base, the more compactly a large number can be written. The decimal number 1,000 requires ten digits in binary notation (1111101000) and 1,000 digits in the 1-system. On the debit side, a large base means more digits to memorize and larger tables of addition and multiplication.

From time to time reform groups, fired with almost religious zeal, seek to overthrow what has been called the "tyranny of 10" and replace it with what they believe to be a more efficient radix. In recent years the duodecimal system, based on 12, has been the most popular. Its chief advantage is that all multiples of the base can be evenly halved, thirded, and quartered. (The unending decimal fraction .3333 . . . , which stands for 1/3, becomes a simple .4 in the 12-system.) There have been advocates of a 12-base since the sixteenth century, including such personages as Herbert Spencer, John Quincy Adams, and George Bernard Shaw. H. G. Wells has the system adopted before the year 2100 in his novel *When the Sleeper Wakes*. There is even a Duodecimal Society of America. (Its headquarters are at 20 Carleton Place, Staten Island, New York 10304.) It publishes *The Duodecimal Bulletin* and *Manual of the Dozen System* and supplies its "dozeners" with a slide rule based on a radix of 12. The society uses an X symbol (called dek) for 10 and an inverted 3 (called el) for 11. The first three powers of 12 are do, gro, mo; thus the duodecimal number 111X is called mo gro do dek.

Advocates of radix 16 have produced the funniest literature. In 1862 John W. Nystrom published privately in Philadelphia his *Project of a New System of Arithmetic, Weight, Measure, and Coins, Proposed to Be Called the Tonal System, with Sixteen to the Base*. Nystrom urges that numbers 1 through 16 be called an, de, ti, go, su, by, ra, me, ni, ko, hu, vy, la, po, fy, ton. Joseph Bowden, who was a mathematician at Adelphi College, also considered 16 the best radix but preferred to keep the familiar names for numbers 1 through 12, then continue with thrun, fron, feen, wunty. In Bowden's notation 255 is written Ɔ̄Ɔ̄ and

pronounced "feenty feen." (See Chapter 2 of his *Special Topics in Theoretical Arithmetic*, privately published; Garden City, New York: 1936.)

It seems unlikely that the "tyranny of 10" will soon be toppled, but that does not prevent the mathematician from using whatever number system he finds most useful for a given task. If a structure is rich in two values, such as the on-off values of computer circuits, the binary system may be much more efficient than the decimal system. Similarly, the ternary, or 3-base, system is often the most efficient way to analyze structures rich in three values. In the quotation that opens this chapter Carson McCullers is writing about herself. She is the woman singing in the darkness about that grotesque triangle in which Macy loves Miss Amelia, who loves Cousin Lymon, who loves Macy. To a mathematician this sad, endless round of unrequited love suggests the endless round of a base-3 arithmetic: each note ahead of another, like the numbers on an eternally running three-hour clock.

In ternary arithmetic the three notes are 0, 1, 2. As you move left along a ternary number, each digit stands for a multiple of a higher power of 3. In the ternary number 102, for example, the 2 stands for 2×1. The 0 is a "place holder," telling us that no multiples of 3 are indicated. The 1 stands for 1×9. We sum these values, $2 + 0 + 9$, to obtain 11, the decimal equivalent of the ternary number 102. Figure 80 shows the ternary equivalents of the decimal numbers 1 through 27. (A Chinese abacus, by the

DECIMAL NUMBERS	TERNARY NUMBERS			
	3^3	3^2	3^1	3^0
1				1
2				2
3			1	0
4			1	1
5			1	2
6			2	0
7			2	1
8			2	2
9		1	0	0
10		1	0	1
11		1	0	2
12		1	1	0
13		1	1	1
14		1	1	2
15		1	2	0
16		1	2	1
17		1	2	2
18		2	0	0
19		2	0	1
20		2	0	2
21		2	1	0
22		2	1	1
23		2	1	2
24		2	2	0
25		2	2	1
26		2	2	2
27	1,	0	0	0

80. Ternary numbers 1 through 27

way, is easily adapted to the ternary system. Just turn it upside down and use the two-bead section.)

Perhaps the most common situation lending itself to ternary analysis is provided by the three values of a balance scale: either one pan goes down or the other pan goes down, or the pans balance. As far back as 1624, in the second edition of a book on recreational mathematics, Claude Gasper Bachet asked for the smallest number of weights needed for weighing any object with an integral weight of from 1 through 40 pounds. If the weights are restricted to one side of the scale, the answer is six: 1, 2, 4, 8, 16, 32 (successive powers of 2). If the weights may go on either pan, only four are needed: 1, 3, 9, 27 (successive powers of 3).

To determine how weights are placed to weigh an object of n pounds, we first write n in the ternary system. Next we change the form of the ternary number so that instead of expressing its value with the symbols 0, 1, 2 we use the symbols, 0, 1, −1. To do this each 2 is changed to −1, then the digit to the immediate left is increased by 1. If this produces a new 2, it is eliminated in the same way. If the procedure creates a 3, we replace the 3 with 0 and add 1 to the left. For instance, suppose the weight is 25 pounds, or 221 in ternary notation. The first 2 is changed to −1, then 1 is added to the left, forming the number 1 −1 2 1. The remaining 2 is now changed to −1, and 1 is added to the left, making the number 1 0 −1 1. This new ternary number is equivalent to the old one ($27 + 0 − 3 + 1 = 25$), but now it is in a form that tells us how to place the weights. Plus digits indicate weights that go in one pan, minus digits indicate weights that go in the other pan. The object to be weighed is placed on the minus side. Figure 81 shows how the three weights are

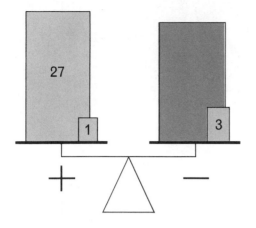

81. How to weigh a 25-pound object

A more sophisticated balance-scale problem (dozens of papers have discussed it since it first sprang up, seemingly out of nowhere, in 1945) is the problem of the 12 coins. They are exactly alike except for one counterfeit, which weighs a bit more or a bit less than the others. With a balance scale and *no* weights, is it possible to identify the counterfeit in three weighings and also know if it is underweight or overweight?

Although I constantly receive letters asking about this problem, I have avoided writing about it because it was so ably discussed by C. L. Stong in "The Amateur Scientist" column of *Scientific American* for May, 1955. Now we shall see how one solution (there are many others) is linked with the ternary system.

First, list the ternary numbers from 1 through 12. To the right of each number write a second ternary number obtained from the first by changing each 0 to 2, each 2 to 0 [*see Figure 82*]. Next, find every num-

placed for weighing a 25-pound object.

The base-3 system using the symbols −1, 0, +1 is called the "balanced ternary system." A good discussion of it can be found in Donald E. Knuth's *Seminumerical Algorithms* (New York: Addison-Wesley,1969; pages 173–175). "So far no substantial application of balanced ternary notation has been made," Knuth concludes, "but perhaps its symmetric properties and simple arithmetic will prove to be quite important some day (when the 'flip-flop' is replaced by a 'flip-flap-flop')."

Suppose you wish to determine the weight of a single object known to have an integral weight of from 1 through 27 pounds. What is the smallest number of weights needed, assuming that they may be placed on either pan? There is no catch, but the question is tricky and the answer is not what you are first likely to think.

82. Ternary numbers for 12-coin problem

1	001	221
2	002	220
3	010	212
4	011	211
5	012	210
6	020	202
7	021	201
8	022	200
9	100	122
10	101	121
11	102	120
12	110	112

ber that contains as the first *unlike* digits one of the following pairs of adjacent digits: 01, 12, 20. Assign one of these 12 numbers (shown in color) to each of the 12 coins.

For the first weighing the four coins with a first digit of 0 go left, the four with a first digit of 2 go right. If the pans balance, put down 1 as the first digit of the counterfeit. If the left pan goes down, the counterfeit's first digit is 0; if the right pan goes down, it is 2.

For the second weighing the four coins with a middle digit of 0 go left, the four with a middle digit of 2 go right. The same procedure is followed to obtain the middle digit of the counterfeit. On the third weighing, coins with final digits of 0 go left, those with final digits of 2 go right, and the last digit of the counterfeit is obtained as before. Figure 83 shows the three weighings that identify the counterfeit as coin 201. When the coin is overweight, as in this case, the number given by the three weighings is the actual number of the coin. If the three weighings give a number *not* assigned to a coin, then the coin is *underweight*. Its number is obtained by substituting a 0 for each 2, and a 2 for each 0.

Scores of simplified versions of this procedure have been devised. The best I know comes from W. Fitch Cheney, Jr., a mathematician at the University of Hartford. Label the coins with the letters of SILENT COWARD. The three weighings are SCAN

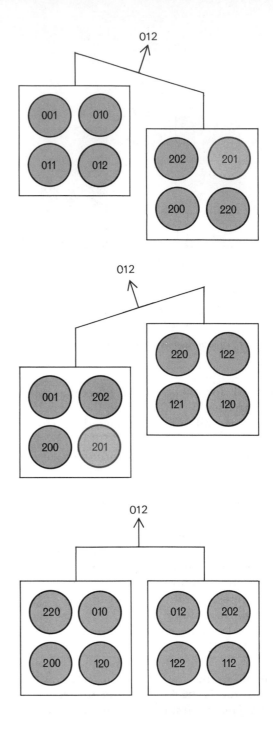

83. *Three weighings to identify a counterfeit coin*

against WORD, SCAR against LINE, SLOT against RAID. Put a ring around each word that goes down. If a pair balances, mark out all its letters from all six words. Inspect the circled words. If there is a letter not crossed out that appears in each word, it indicates the false coin and the coin is overweight. If there is no such letter, you are sure to find one not crossed out in each of the uncircled words. It then indicates an underweight counterfeit. Other key words can, of course, be devised. L. E. Card, intrigued by Cheney's SILENT COWARD, found two dozen sets, of which I cite only one: CRAZY WEIGHTS: CITY-HAZE, GREW-HAZY, AND WISH-TRAY.

The problem has been generalized. In four weighings one can identify the false coin, and tell whether it is light or heavy, among a maximum of $3^1 + 3^2 + 3^3 = 39$ coins; five weighings will take care of $3^1 + 3^2 + 3^3 + 3^4 = 120$ coins, and so on. More compactly, n weighings take care of $\frac{1}{2}(3^n - 3)$ coins. It is worth noting that a counterfeit among 13 coins can be found in three weighings if one need not know whether it is heavier or lighter (simply put the 13th coin aside and if you fail to find the counterfeit among the 12, the 13th coin is it); to know whether the false coin is heavier or lighter, three weighings also suffice for 13 coins if you add a 14th coin known to be genuine.

Many card tricks are closely related to the 12-coin problem. One of the best is known as Gergonne's three-pile problem after the French mathematician Joseph Diez Gergonne, who first studied it early in the

nineteenth century. Someone is asked to look through a packet of 27 cards and fix one in his mind. He holds the packet face down, deals the cards face up into a row of three, then continues dealing on top of these cards, left to right, until all 27 have been dealt into three face-up piles of nine cards each. After telling the magician which pile contains his chosen card, he assembles the piles by placing them on top of one another, in any order he wishes, turns the packet face down and again deals them into three face-up piles. Once more he indicates the pile in which his card fell. This is repeated a third time, then the assembled packet is placed face down on the table. The magician, who has not touched the cards throughout the entire procedure, names the position of the chosen card.

The secret lies in observing, at each pickup, whether the pile with the selected card goes on the top, the bottom, or in the middle of the assembled facedown packet. These positions are designated 0 for the top, 1 for the middle, 2 for the bottom. The ternary number expressed by the three pickups, written from *right to left*, is the number of cards above the chosen card after the final pickup. For example, suppose the first pickup puts the pile on the top (0), the second on the bottom (2), the third in the middle (1). These digits, written right to left, give the ternary number 120, or 15 in the decimal scale. Fifteen cards are therefore above the selected one, making it the 16th card from the top. Of course, the trick can be done just as easily in reverse. The spectator chooses any number from 1 through 27, then the magician, making the pickups himself, brings the card to that number from the top.

If in dealing into three piles one is permitted to put each card on *any* pile, a powerful sorting method results. At this point the reader is asked to obtain eight file cards and print on each card one of the letters in the word DEMOCRAT. Arrange the cards into a packet, letter sides down, that spells DEMOCRAT from the top down [*see top illustration of Figure 84*]. You wish to rearrange the cards so that, from the top down, they are in alphabetical order as shown in Figure 84, bottom. It is easily done in one deal. Turn the top card, *D*, face up and place it as the first card of pile 1. The next three cards, *E, M, O*, go on top of the *D*. *C* becomes the first card of pile 2, *R* goes back on pile 1, *A* starts pile 3, and *T* goes on pile 1. Assemble by putting pile 1 on 2 and those cards on 3; then turn the

packet face down. You will find the cards in alphabetical order, top to bottom. A single deal is also sufficient, as you can easily discover, for changing the alphabetized order back to DEMOCRAT.

Put the DEMOCRAT cards aside and make a new set that spells REPUBLICAN. Can *this* set be alphabetized in one operation? No, it cannot. What is the smallest number of operations necessary? Remember, the initial packet of face-down cards must spell the word from the top down. Each card is dealt face up, the piles are picked up in any order, then the packet is turned face down to conclude one operation. After the last operation the cards must be in the order ABCEILNPRU, top to bottom. If you solve this problem, see if you can determine the minimum number of operations needed to change the order back to REPUBLICAN. And if both problems seem too easy, try a set of cards that spell SCIENTIFIC AMERICAN. In

84. *Original* (top) *and desired sorting of DEMOCRAT cards*

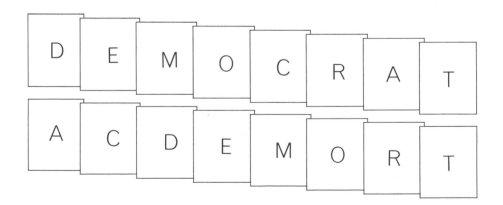

the answer section I explain how all sorting problems of this type can be solved quickly by a simple application of ternary numbers, and I also answer the problem of the weights.

Answers

The minimum number of weights needed to weigh 27 boxes with integral weights of from 1 through 27 pounds, assuming that weights may be placed on either side of a balance scale, is three: 2, 6, and 18 pounds. (They represent doublings of successive powers of 3.) These weights will achieve an exact balance for every even number of pounds from 1 through 27. The odd weights are determined by checking the even weights directly above and below; for example, a box of 17 pounds is identified by the fact that it weighs less than 18 and more than 16 pounds. (Mitchell Weiss of Downey, California, provided this pleasant twist on an old problem.)

The task of alphabetizing the letters of REPUBLICAN by dealing letter cards into three piles can be solved in two operations. First, write down the letters in alphabetical order: ABCEILNPRU. A is the first letter, so we place a 0 above the letter A in the word REPUBLICAN. We move *right* along the word in search of B, the second letter, but we do not find it. Because we are forced to move *left* to reach B, we put 1 above it. We continue to move right in search of C. This time we find it on the right, so we label it with 1 also. The next letter, E, forces us to

move left again, therefore we label it 2. I is to the right of E, so it gets 2 also, but L carries us left again, so it gets 3. In short, we raise the number only when we have to move left to find the letter. This is how the final result appears:

```
5  2  4  5  1  3  2  1  0  3
R  E  P  U  B  L  I  C  A  N
```

On each letter card write the ternary equivalent of the decimal number assigned to that letter. The cards are held in a face-down packet, spelling REPUBLICAN from the top down. Imagine that the three piles are numbered, left to right, 0, 1, 2. Turn over the top card, R. Its ternary number is 12. The *last* digit, 2, tells you to deal the card to pile 2 (the end pile on the right). The next card, E, has a ternary number of 02; it also goes on the right end pile. Continue in this way, dealing each card to the pile indicated by the final digit. The piles are always assembled from right to left by putting the last pile (2) on the center pile (1), then all those cards on the first pile (0). Turn the packet face down and deal once more, this time dealing as indicated by the *first* digits of each ternary number. Assemble as before. The cards are now alphabetized.

To put the cards back in their original order a new analysis of the letters must be made, assigning them a new set of numbers:

```
5  2  4  1  3  2  5  1  0  1
A  B  C  E  I  L  N  P  R  U
```

Two operations will return the cards to

their initial order, but the sorting procedure is not the same as before. If the decimal numbers assigned to the letters go above 8, then a ternary number for a letter will require more than two digits, and the number of required operations will be more than two. It is easy to see that the minimum number of operations is given by the number of digits in the highest ternary number. To alphabetize SCIENTIFIC AMERICAN the letters are numbered:

```
6 1 4 2 5 6 4 3 3 1
S C I E N T I F I C
  0 4 2 5 3 1 0 4
  A M E R I C A N
```

Because the highest number, 6, has only two digits in its ternary form, only two operations are called for. However, to reverse the procedure, changing the alphabetized order back to SCIENTIFIC AMERICAN, the highest number is 10. This has three ternary digits,

therefore three operations are necessary. If the reader will test the system on longer phrases or sentences, he will be astonished at how few operations are required for what seems to be an enormously difficult sorting job. One can generalize the method to any number of piles, n, simply by writing numbers in a system based on n.

References

"The Problem of the Pennies." F. J. Dyson. *The Mathematical Gazette,* Vol. 30, No. 291; October, 1946. Pages 231–234.

"The Counterfeit Coin Problem." C. A. B. Smith. *The Mathematical Gazette,* Vol. 31, No. 293; February, 1947. Pages 31–39.

"On Various Versions of the Defective Coin Problem." Richard Bellman and Brian Gluss. *Information and Control,* Vol. 4, Nos. 2–3; September, 1961. Pages 118–131.

Puzzles and Paradoxes. T. H. O'Beirne. New York: Oxford University Press, 1965. Chapters 2 and 3.

12. The Trip around the Moon and Seven Other Problems

1. The Trip around the Moon

The year is 1984. A moon base has been established and an astronaut is to make an exploratory trip around the moon. Starting at the base, he is to follow a great circle and return to the base from the other side. The trip is to be made in a car built to travel over the satellite's surface and having a fuel tank that holds just enough fuel to take the car a fifth of the way around the moon. In addition the car can carry one sealed container that holds the same amount of fuel as the tank. This may be opened and used to fill the tank or it may be deposited, unopened, on the moon's surface. No fraction of the container's contents may be so deposited.

The problem is to devise a way of making the round trip with a minimum consumption of fuel. As many preliminary trips as desired may be made, in either direction, to leave containers at strategic spots where they can be picked up and used later, but eventually a complete circuit must be made all the way around in one direction. Assume that there is an unlimited supply of containers at the base. The car can always be refueled at the base from a large tank. For example, if it arrives at the base with a partly empty tank, it can refill its tank without wasting the fuel remaining in its tank.

To work on the problem, it is convenient to draw a circle and divide it into twentieths as shown on the next page in Figure 85. Fuel used in preliminary trips must of course be counted as part of the total amount consumed. For example, if the car carried a container to point 90, left it there and returned to base, the operation would consume one tank of fuel.

This operations-research problem is similar in some respects to the well-known problem of crossing a desert in a truck, but it demands a quite different analysis.

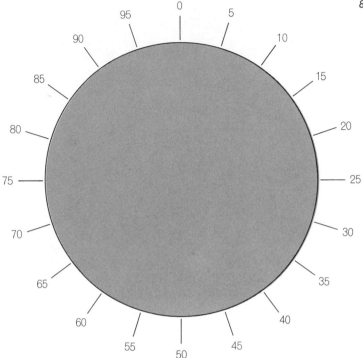

2. The Rectangle and the Oil Well

An oil well being drilled in flat prairie country struck pay sand at an underground spot exactly 21,000 feet from one corner of a rectangular plot of farmland, 18,000 feet from the opposite corner, and 6,000 feet from a third corner. How far is the underground spot from the fourth corner? Readers who solve the problem will discover a useful formula of great generality and delightful simplicity.

3. Wild Ticktacktoe

A. K. Austin of Hull, England, has written to suggest a wild variation of ticktacktoe. It is the same as the standard game except that each player, at each turn, may mark either a naught or a cross. The first player to complete a row of three (either three naughts or three crosses) wins the game.

Standard ticktacktoe is a draw if both sides play rationally. This is not true of the unusual variant just described. Assuming that both players adopt their best strategy, who is sure to win: the first or the second player?

4. Coins of the Realm

In this country at least eight coins are required to make the sum of 99 cents: a half-dollar, a quarter, two dimes and four pennies. Imagine yourself the leader of a small, newly independent nation. You have the task of setting up a system of coinage based on the cent as the smallest unit. Your objective is to issue the smallest number of different coins that will enable any value from 1 to 100 cents (inclusive) to be made with no more than two coins.

For example, the objective is easily met with 18 coins of the following values: 1, 2, 3, 4, 5, 6, 7, 8, 9, 10, 20, 30, 40, 50, 60, 70, 80, 90. Can the reader do better? Every value must be obtainable either by one coin or as the sum of two coins. The two coins need not, of course, have different values.

5. Bills and Two Hats

"No," said the mathematician to his 14-year-old son, "I do *not* feel inclined to increase your allowance this week by ten dollars. But if you'll take a risk, I'll make you a sporting proposition."

The boy groaned. "What is it this time, Dad?"

"I happen to have," said his father, "ten crisp new ten-dollar bills and ten crisp new one-dollar bills. You may divide them any way you please into two sets. We'll put one set into hat *A*, the other set into hat *B*. Then I'll blindfold you. I'll mix the contents of each hat and put one hat on the right and one on the left side of the mantel. You pick either hat at random, then reach into that hat and take out one bill. If it's a ten, you may keep it."

"And if it isn't?"

"You'll mow the lawn for a month, with no complaints."

The boy agreed. How should he divide the 20 bills between the two hats in order to maximize the probability of his drawing a ten-dollar bill, and what will that probability be?

6. Dudeney's Word Square

Charles Dunning, Jr., of Baltimore, Maryland, recently set himself the curious task of placing letters in the nine cells of a three-by-three matrix so as to form the largest possible number of three-letter words. The words may be read from left to right or right to left, up or down and in either direction along each of the two main diagonals. Dunning's best result, shown in Figure 86, gives ten words: tea, urn, bay, tub, but, era, are, any, try, bra.

How close it is possible to come in English to the theoretical maximum of 16 words? A letter may be used more than once, but words must be different in order to count. They should be dictionary words.

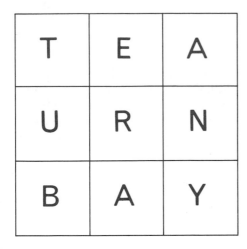

T	E	A
U	R	N
B	A	Y

86. Dunning's 10-word square

I have on hand a specimen from one of H. E. Dudeney's puzzle books that raises the number of words to 12 but perhaps readers can do better.

7. Ranking Weights

Five objects, no two the same weight, are to be ranked in order of increasing weight. You have available a balance scale but no weights. How can you rank the objects correctly in no more than seven separate weighings?

For two objects, of course, only one weighing is required. Three objects call for three weighings. The first determines that A is heavier than B. We then weigh B against C. If B is heavier, we have solved the problem in two weighings, but if C is heavier, a third weighing is required to compare C with A. Four objects can be ranked easily with no more than five weighings.

With five objects the problem ceases to be trivial. As far as I know, no general method for ranking *n* objects with a minimum number of weighings has yet been established.

8. Queen's Tours

Hundreds of entertaining puzzles, known as "chess tours," involve the movements of single chess pieces over the board. Chapter 10 of this book discussed knight's tours and their connection with graph theory. Here is a choice selection of five queen's-tour problems. The reader does not have to be a chess player to work them out; he need only know that the queen moves an unlimited distance horizontally, vertically, or diagonally. The problems are roughly in order of increasing difficulty.

1. Place the queen on square A [*see Figure* 87]. In four continuous moves traverse all nine of the gray-shaded squares.

2. Place the queen on cell D (the white queen's starting square) and make the longest trip possible in five moves. ("Longest" means the actual length of the path, not the maximum number of cells traversed.) The queen must not visit the same cell twice, and she is not allowed to cross her own path. Assume the path to be through the center points of all cells.

3. Place the queen on cell B. In 15 moves pass through every square once and only once, ending the tour on cell C.

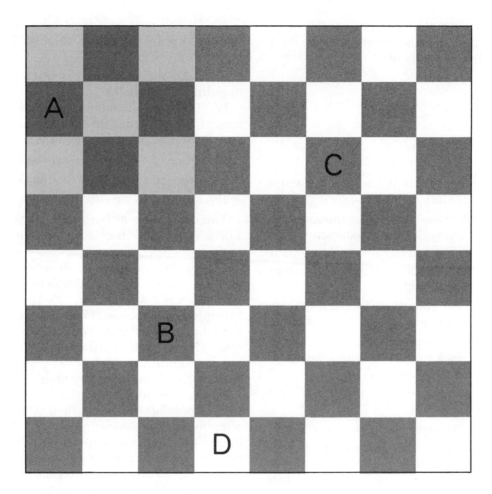

87. Board for queen's-tour problems

4. Place the queen on a corner square. In 14 moves traverse every cell of the board, returning to the starting square on the 14th move. Individual cells may be visited more than once. This "reentrant queen's tour" was first published in 1867 by Sam Loyd, who always considered it one of his finer achievements. The tour, whether reentrant or open at the ends, cannot be made in fewer than 14 moves.

5. Find a similar reentrant tour in 12 moves on a seven-by-seven board. That is,

the queen must start and end on the same cell and pass through every cell at least once. As before, cells may be entered more than once.

Answers

1.

The moon can be circled with a consumption of 23 tankfuls of fuel.

1. In five trips, take five containers to point 90, return to base (consumes five tanks).

2. Take one container to point 85, return to point 90 (one tank).

3. Take one container to point 80, return to point 90 (one tank).

4. Take one container to point 80, return to point 85, pick up the container there and take it to point 80 (one tank).

5. Take one container to point 70, return to point 80 (one tank).

6. Return to base (one tank).

This completes all preliminary trips in the reverse direction. There is now one container at point 70, one at point 90. Ten tanks have been consumed.

7. Take one container to point 5, return to base (half a tank).

8. In four trips, take four containers to point 10, return to base (four tanks).

9. Take one container to point 10, return to point 5, pick up the container there and leave it at point 10 (one tank).

10. In the next two trips take two containers to point 20, return to point 10 (two tanks).

11. Take one container to point 25, return to point 20 (one tank).

12. Take one container to point 30, return to point 25, pick up the container there and carry it to point 30 (one tank).

13. Proceed to point 70 (two tanks).

14. Proceed to point 90 (one tank).

15. Proceed to base (half a tank).

The car arrives at base with its tank half filled. The total fuel consumption is 23 tanks.

I found this problem, in the story form of circling a mountain, as Problem 50 in H. E. Dudeney's *Modern Puzzles* (1926); it is reprinted as Problem 77 in my edition of Dudeney's *536 Puzzles and Curious Problems* (New York: Scribner's, 1967). The above solution, which was supplied in variant forms by many readers, is essentially the same as Dudeney's.

It is not, however, minimal. Wilfred H. Shepherd, Manchester, England, first reduced the fuel consumption to $22^7/_{12}$ containers. This was further reduced by Robert L. Elgin, Altadena, California, to $22^9/_{16}$ containers. His solution can be varied in trivial ways, but it is believed to be minimal.

Elgin's solution is best explained by dividing the circle into 80 equal parts. A container is picked up every time you move away from home base, and left on the ground every time you turn to move toward the base. Any available container is emptied into the car's tank every time the car runs out of fuel. Assume that each container holds $^{80}/_5 = 16$ units of fuel. The solution follows:

1. Take one container to point 73, return to base (consumes 14 tanks).

2. Two containers to 75, return to base (20 tanks).

3. Two containers to 72, return to base (32 tanks).

4. One container to 69½, back to 75 (16 tanks).

5. One container to 67½, back to 69½, forward to 67½, back to 72 (16 tanks).

6. One container to 64, back to 67½, forward to 66, back to 67½, forward to 66 (16 tanks).

7. One container to 57, back to 64 (16 tanks).

8. Return to base (16 tanks).

9. Five containers to 8, return to base (80 tanks).

10. One container to 10, back to 8, forward to 10, back to 8 (16 tanks).

11. One container to 16, back to 8 (16 tanks).

12. One container to 16½, back to 16, forward to 16½, back to 10 (16 tanks).

13. One container to 21¼, back to 16½ (16 tanks).

14. One container to 25, back to 21¼, forward to 25 (16 tanks).

15. One container to 41 (16 tanks).

16. Proceed to 57 (16 tanks).

17. Proceed to 73 (16 tanks).

18. Proceed to base (7 tanks).

Total fuel consumption is $^{361}/_{16} = 22^9/_{16}$ tanks.

2.

Consider first a spot p on the surface inside the rectangle shown at the top of Figure 88. Adding two broken coordinate lines provides a set of right triangles. Because $e^2 = a^2 + c^2$ and $g^2 = b^2 + d^2$, we can write the equality

$$e^2 + g^2 = a^2 + c^2 + b^2 + d^2.$$

And since $f^2 = a^2 + d^2$ and $h^2 = b^2 + c^2$, we can write

$$f^2 + h^2 = a^2 + d^2 + b^2 + c^2.$$

The right sides of both equations are the same, therefore

$$e^2 + g^2 = f^2 + h^2.$$

Exactly the same analysis applies to the bottom diagram, in which point p is outside the rectangle. If you think of p in either diagram as belowground, this will lengthen certain sides of the right triangles involved, but the relations expressed by the equations remain unchanged. In other words, regardless of where point p is located in space — above, below or even on the edge or corner of the rectangle itself — the sum of the squares of its distances from two opposite corners of the rectangle will equal the sum

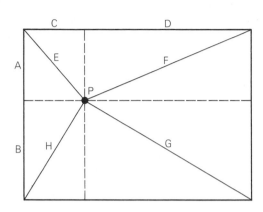

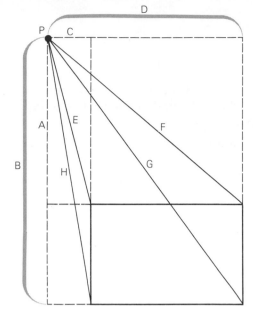

88. *Solution to the oil-well problem*

of the squares of its distances from the other two corners. Applying this simple formula to the three distances given yields 27,000 as the fourth distance. The sides of the rectangle are not, of course, determined by the given data.

3.

When ticktacktoe players are allowed to play either a naught or a cross on each move, the first player can always win by first taking the center cell. Suppose he plays a cross. The second player has a choice of marking either a corner or a side cell.

Assume that he marks a corner cell. To avoid losing on the next move he must mark it with a 0. The first player replies by putting a 0 in the opposite corner, as in diagram

a in Figure 89. The second player cannot prevent his opponent from winning on his next move.

What if the second player takes a side cell on his second move? Again he must use a 0 to avoid losing on the next move. The first player replies as shown in the next diagram [*b*]. The second player's next move is forced [*c*]. The first player responds as shown in the final diagram [*d*], using either symbol. Regardless of where the second player now plays, the first player wins on his next move.

Wild ticktacktoe (as this game was called by S. W. Golomb) immediately suggests a variant: reverse wild ticktacktoe. The rules are as before except that the first player to get three like-symbols in a row *loses*. Robert Abbott was the first to supply

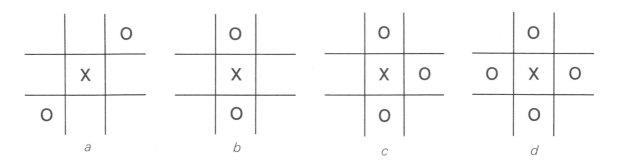

89. The ticktacktoe problem

a proof that the game is a draw when played rationally. The first player cannot assure himself a win, but can always tie by using a symmetry strategy similar to his strategy for obtaining a draw in ordinary (or "tame") reverse ticktacktoe. He first plays any symbol in the center. Thereafter he plays symmetrically opposite the second player, always choosing a symbol different from the one previously played.

As Golomb has pointed out, this strategy gives at least a draw in reverse ticktacktoe (wild or not) on all boards of odd-order. On even-order boards the second player can obtain at least a draw by a similar strategy. With the order-3 cubical board, Golomb adds, on which a draw is impossible, the strategy assures a win for the first player in reverse ticktacktoe, wild or tame.

4.

With as few as 16 different coins one can express any value from 1 cent to 100 cents as the sum of no more than 2 coins. The coins are: 1, 3, 4, 9, 11, 16, 20, 25, 30, 34, 39, 41, 46, 47, 49, 50. This solution is given, without proof that it is minimal, in Problem 19 of Roland Sprague's *Recreation in Mathematics*, translated from the German by T. H. O'Beirne (London: Blackie and Son, 1963).

Sprague's solution has a range of only 100. A 16-integer solution with the higher range of 104 was provided by Peter Wegner of the University of London: 1, 3, 4, 5, 8, 14, 20, 26, 32, 38, 44, 47, 48, 49, 51, 52.

5.

The boy maximizes his chance of drawing a ten-dollar bill by putting a single ten-dollar bill in one hat, the other 19 bills (9 ten-dollar bills and 10 one-dollar bills) in the other hat. His chance of picking the hat with the ten-dollar bill is 1 in 2, and the probability of picking a ten-dollar bill from that hat is 1 (certain). If he picks the other hat, there is still a probability of 9/19 that he will draw a ten-dollar bill from it.

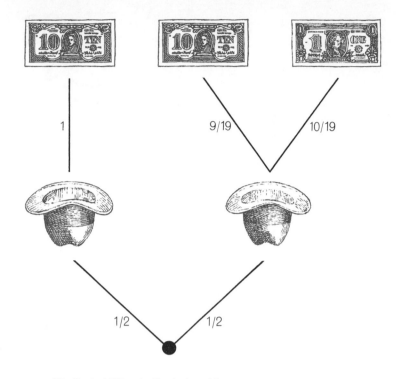

90. Probabilities in the hat problem

This simple stochastic process is shown in Figure 90. The probability that he will draw a ten-dollar bill from hat A is $1/2 \times 1$, or 1/2. The probability that he will draw a ten-dollar bill from hat B is $1/2 \times 9/19$, or 9/38. The sum of the two probabilities, 14/19 (or almost 3/4), is his overall probability of getting a ten-dollar bill.

6.

Dudeney, in his posthumously published *A Puzzle-Mine*, was able to achieve 12 good English words by placing letters on the 9-cell square like this:

G E T

A I A

S U P

The words are: get, teg, sup, pus, pat, tap, gas, sag, pig, gip, sit, aia. If the contraction "'tis" is permitted, the number is 13.

More than 50 readers sent in 12-word squares, most of them superior to Dudeney's 12-worder. Many readers showed how 12 words could be obtained from a cross of A's in the center of the square, as shown [*number 1*] in Figure 91. Twenty-six readers sent in 13-word squares, in most cases with words that could all be found in *Webster's New Collegiate Dictionary*. The typical square [*numbered 2 in the illustration*] was independently discovered by Vaughn Baker, Mrs. Frank H. Driggs, William Knowles, and Alfred Vasko.

Vaughn Baker, David Grannis, Horace Levinson, H. P. Luhn, Stephen C. Root, Hugh Rose, Frank Tysver, C. Brooke Worth, and George Zinsmeister all produced 14-word squares. Baker's square [*numbered 3 in the illustration*] has only one word— "wey"—that is not usually found in short

dictionaries. Frederick Chait, James Garrels, B. W. Le Tourneau, Marvin Weingast, and Arnold Zeiske devised 15-worders, but none with more than 12 short dictionary words.

Five readers hit the jackpot with 16 words: Dmitri Borgmann, L. E. Card, Mrs. D. Harold Johnson, Peter Kugel, and Wylie Wilson. The five squares [*numbered 4 through 8*] are reproduced in the order in which the alphabetized names appear above. There is no way to decide which square is best, since all exploit obscure words and even the meaning of "word" is hazy.

Several readers experimented with order-4 squares. L. E. Card, of Urbana,

91. *Solutions to word-square puzzle:* (1) *12-word,* (2) *13-word,* (3) *14-word,* (4–8) *16-word*

M	A	R
A	A	A
P	A	T

1

P	I	G
E	A	U
R	O	T

2

S	T	Y
U	A	E
P	O	W

3

A	T	E
R	A	E
T	O	R

4

E	R	A
L	E	E
S	A	N

5

A	T	S
R	I	A
T	A	D

6

S	E	R
T	A	O
A	R	D

7

E	E	L
T	A	O
A	R	T

8

Illinois, achieved the maximum (20 dictionary words), with:

S N A P

A E R A

R A I L

T R A P

"Tras" is the plural of "tra," a Malaysian coin.

7.

Five objects can be ranked according to weight with no more than seven weighings on a balance scale:

1. Weigh A against B. Assume that B is heavier.

2. Weigh C against D. Assume that D is heavier.

3. Weigh B against D. Assume that D is heavier. We now have ranked three objects: $D > B > C$.

4. Weigh E against B.

5. If E is heavier than B, we now weigh it against D. If E is lighter than B, we weigh it against A. In either case E is brought into the series so that we obtain a rank order of four objects. Assume that the order is $D > B > E > A$. We already know (from step 2) how the remaining object C compares with D. Therefore we have only to find C's place with respect to the rank order of the other three. This can always be done in two weighings. In this case:

6. Weigh C against E.

7. If C is heavier than E, weigh it against B. If C is lighter than E, weigh it against A.

The general problem of ranking n weights with a minimum number of weighings (or n tournament players with a minimum number of no-draw two-person contests) was first proposed by Hugo Steinhaus. He discusses it briefly in the 1950 edition of *Mathematical Snapshots* and includes it as Problem 52 (with $n = 5$) in *One Hundred Problems in Elementary Mathematics* (New York: Basic Books, 1964). In the latest revision of *Mathematical Snapshots* (New York: Oxford University Press, 1968), Steinhaus gives a formula that provides correct answers through $n = 11$. (For 1 through 11 the minimum number of weighings are 0, 1, 3, 5, 7, 10, 13, 16, 19, 22, 26.) The formula predicts 29 weighings for 12 objects, but it has been proved that the minimum number is 30.

The general problem is discussed by Lester R. Ford and Selmer M. Johnson, both of the Rand Corporation, in "A Tournament Problem," in *The American Mathematical Monthly*; May, 1969. For a more recent discussion of this and closely related problems, see Section 5–3–1 of Donald Knuth's *Sorting and Searching* (New York: Addison-Wesley, 1971).

8.

Answers to the five queen's-tour problems are shown in Figure 92. In the fourth and fifth problems there are solutions other than those shown, but none in fewer moves.

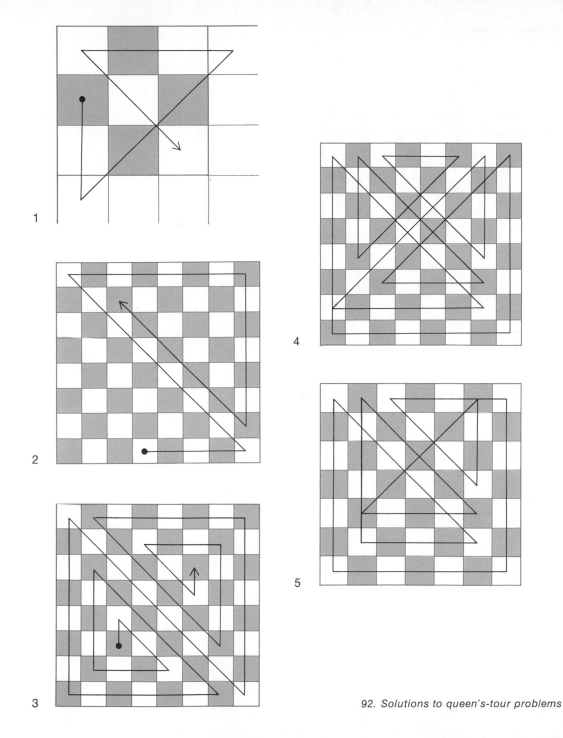

1

2

3

4

5

92. Solutions to queen's-tour problems

If you solved the third problem by moving first to the lower right corner, up to the upper right corner, along a main diagonal to the lower left corner, up to the upper left and then right seven squares, you found a path almost (but not quite) as long as the one shown.

13. The Cycloid: Helen of Geometry

DO THE TOPS of the tires on a moving car go faster than the bottoms? This odd question will start as many ferocious parlor debates as the old problem about the man who walks around a tree trying to see a squirrel on the opposite side of the trunk. As he walks, the squirrel scurries around the tree, keeping its belly against the trunk so that it always faces the man but with the trunk constantly hiding it from view. When the man has circled the tree, has he also gone around the squirrel?

William James, considering this weighty metaphysical problem in the second chapter of his book *Pragmatism*, concludes that it all depends on what one means by "around." Similarly, the tire question cannot be answered without prior agreement as to precisely what all the words mean. Let us say that by "top" and "bottom" of the tire we mean those points on the tire that are at any given moment close to the top or bottom, and that by "go faster" we refer to the horizontal velocity of those points

in relation to the ground. Surprising as it may seem, points near the top do move faster than points near the bottom.

This can be demonstrated by a simple experiment with a coffee can. Cover the bottom of the can with white paper. Using a dark crayon, draw about eight diameters, like the spokes of a wheel, on the circular sheet. Place the can on its side and roll it back and forth past your line of vision. Do *not* follow the can with your eyes; keep your gaze fixed on a distant object so that your eyes do not move as the can rolls by. You will find that the black spokes are visible only in the lower half of the wheel. The upper half is a gray blur. The reason is that the spokes in the upper half are actually moving past your eyes at a much faster rate than the spokes in the lower half. This was such a familiar phenomenon in horse-and-buggy days that artists often indicated the motion of wheels by showing distinct spokes only below the axles.

Figure 93 traces the motion of a point on

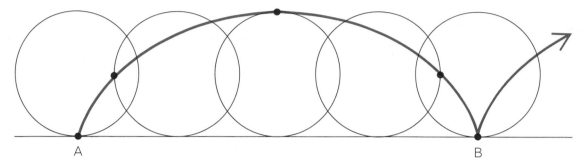

93. *How a cycloid is generated by a point on a rolling circle*

the circumference of a circle as it rolls without slipping along a horizontal line for a distance *AB* that is equal to the circumference of the circle. The position of the circle is shown after each quarter-turn. Assume that the circle rolls with uniform speed. It is easy to see that the point is motionless for an instant on the ground at *A*, gradually increases in speed, reaches its maximum at the highest spot and then accelerates negatively until it touches ground again at *B*. If the wheel continues to roll, the point will trace a series of arches, coming to rest for an instant at the bottom of each cusp. The velocity of the point along the curve conforms to what physicists call a simple harmonic motion. On wheels that have flanges, such as the wheels of a train, points on the flange actually move *backward* while they execute a tiny loop below the level of the track.

The generic name for a curve traced by a point on any type of curve when it rolls

without slipping along any other type of curve is "roulette." In this case a circle rolls on a straight line to generate one of the simplest of roulettes, the cycloid. It has been called the "Helen of geometry," not only because of its beautiful properties but also because it has been the object of so many historic quarrels between eminent mathematicians.

No one knows who first recognized the cycloid as a curve worth studying. There is no mention of it before 1500. The first important treatise on the curve was written in 1644 by the Italian physicist Evangelista Torricelli, a student of Galileo's. Fourteen years later Blaise Pascal, who had abandoned mathematics for a life of religious contemplation, found himself suffering from a terrible toothache. To take his mind off the pain he began thinking about the cycloid. The pain stopped. Regarding this as a sign that God was not displeased with his thoughts, Pascal spent the next eight days in furious research on the curve. His remarkable results were issued first as a series of challenges to other mathematicians and then as a treatise on the cycloid.

One of the simplest questions to ask about the cycloid—although by no means the easiest to answer—is: How long is it? Assume that the generating circle has a diameter of 1. The base line AB will, of course, be pi, an irrational number. Everyone expected the length of the curve to be irrational also. Sir Christopher Wren, the distinguished English architect, apparently was the first to show (in 1658) that the length of the cycloidal arch, from cusp to cusp, is precisely four times the diameter of the circle.

The area below the arch had been measured previously and it too had been a surprise. Galileo had guessed the area to be pi times the area of the generating circle, an estimate obtained by the direct method of cutting the arch from thin material and comparing its weight with that of the circle cut from the same material. Torricelli astounded his colleagues in Italy by proving that the area under the arch is exactly three times the area of the circle. Actually this had been shown earlier by the French mathematician Gilles Personne de Roberval. Torricelli may or may not have known this. Pascal accused Torricelli of deliberately stealing Roberval's proof, as did Roberval himself. In France, René Descartes insisted that the entire problem was trivial. He worked out a simpler way to find the area and challenged Roberval to construct tangents to the cy-

cloid. This led to a long, bitter dispute between the two men. Today all these problems are solved in first-year calculus classes (where the curve is called the "student's curve" because the answers are so simple), but in the seventeenth century calculus was still primitive.

The mechanical properties of the cycloid are as remarkable as its geometric ones. In high school physics one learns that the time it takes a pendulum to swing back and forth is the same regardless of how wide the swing is, but this is only approximate. When the swings are wide, there are slight deviations. In what path should a pendulum swing so that its period is exactly the same regardless of amplitude? Such a curve, called an isochrone, was first discovered by the Dutch physicist Christian Huygens, who published his discovery in 1673. If we turn two cycloidal arches upside down, as shown in Figure 94, and let a pendulum on a cord swing between them, the pendulum will trace what is called the involute of the cycloid. It turns out that the involute is another cycloid of the same size, and that the cycloidal pendulum is isochronal.

For small swings a circular arc is so nearly the same as the central portion of a cycloid that the circular pendulum is almost isochronal, but if the swings vary even a small amount, the "circular error" is cumulative. For example, if a seconds pendulum has a circular arc of two degrees, an increase to three degrees will cause it to lose about .66 second per day. Huygens constructed a pendulum clock—the first ever made—using a flexible pendulum that swung between

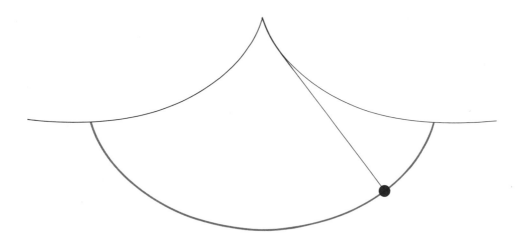

94. *Isochronal pendulum between cycloidal cheeks traces a cycloid*

two cycloidal cheeks. Unfortunately friction on the cheeks produced a greater error than the cycloidal path corrected; clockmakers found it more practical to arrange things so that a circular pendulum would keep a constant amplitude.

It was Huygens who also discovered that the cycloid is the tautochrone, or curve of equal descent. Imagine a marble rolling without friction down an inverted cycloid. No matter where you start it on the curve, it will reach the bottom in the same length of time. (Melville makes reference to this property of the cycloid in an interesting discussion of the structure of whaling ships in Chapter 46 of *Moby Dick*.) Consider a bowl with sides that curve in such a way that any cross section through the center of the bowl will be a cycloid. Marbles placed

at various heights on the sides of the bowl and released simultaneously will reach the center of the bowl at the same instant. Each marble moves with a simple harmonic motion, as does the isochronal pendulum.

The brachistochrone, or curve of *quickest* descent, was not discovered for another score of years. Suppose you are given two points: *A* and *B*. *B* is lower than *A* but not directly below it. The problem is to find a curve connecting *A* and *B* such that a marble, rolling without friction, will travel from *A* to *B* in the shortest possible length of time. This problem was first posed in 1696 by Johann Bernoulli, the Swiss mathematician and physicist, in *Acta Eruditorum*, a famous scientific journal of the day. It was first solved by Johann's brother Jakob (with whom Johann was feuding), but it was also solved by Johann, Leibniz, Newton, and others. Newton solved it, along with a related problem, in 12 hours. (The problem reached him at 4:00 P.M.; he had the solution by 4:00 A.M. and sent it off in the morning.) The brachistochrone turned out to be, as the reader has no doubt guessed, the cycloid. Johann Bernoulli's proof has become a classic of nonrigorous, intuitive reasoning. He found the problem equivalent to one concerning the path of a light ray refracted by transparent layers of steadily decreasing density. The interested reader will find his elegant proof clearly explained in *What Is Mathematics?* by Richard Courant and Herbert Robbins (New York: Oxford University Press, 1941), as well as in Ernst Mach's earlier work, *Science of Mechanics* (Chicago: Open Court Publishing Company, 1893).

Suppose we are given two points, *A* and *B* [*see Figure 95*], and we wish to find the brachistochrone that connects them. What we first find is the radius of the circle that, when rolled against line *AC*, will generate a cycloid starting at *A* and passing through *B*. To do this we place a circle of any size whatever under *AC* and mark a point on its circumference at *A*. The circle is rolled along *AC* until this point crosses *AB*. Assume that it crosses at *D*. Since all cycloids have similar shapes, we know that *AD* is to *AB* as the radius of the large circle we have just used is to the radius of the smaller circle we seek. This smaller circle, rolled along *AC*, will generate a cycloid from *A* to *B*.

Note that in this case the marble actually rolls *uphill* to reach *B*. Nevertheless, it reaches *B* in a shorter time than it would

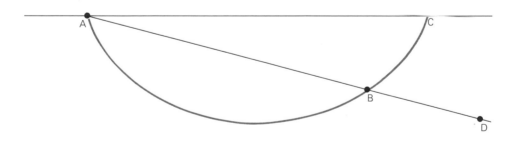

95. *Constructing the curve of quickest descent between* A *and* B

by rolling along a straight line, the arc of a circle or any other curve. Even when A and B are on the same horizontal level, a frictionless marble rolls from A to B in the shortest possible time. (On a straight horizontal line, of course, it would not roll at all.)

An industrious reader should have little difficulty constructing a model for demonstrating the brachistochrone. To draw a large cycloid the coffee can mentioned earlier can be used. A piece of string looped once around it and fastened to the ends of a plank will keep the can from slipping as it rolls along the plank [*see Figure 96*]. A black crayon is taped to the inside of the can so that when the can is rolled along a wall the crayon will trace a cycloid on a

sheet of paper fastened to the wall. Using this trace as a pattern, one can bend stiff wire into a cycloid down which a heavy nut will slide or a double cycloidal track down which a marble will roll. The track can also be formed by the cut edges of two rectangular sheets of plywood or heavy cardboard, mounted vertically, with small strips of wood glued between them to keep the edges separated just enough to carry the marble. Similar tracks should be made to carry a second marble down a circular arc and a third marble down a straight line. The three tracks are placed side by side so that the marbles can be released simultaneously by a pencil held horizontally. (Steel balls can be held by electromagnets and released by pushing a button.) If the

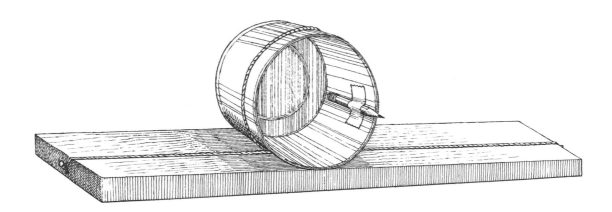

96. Coffee-can device for drawing a cycloid

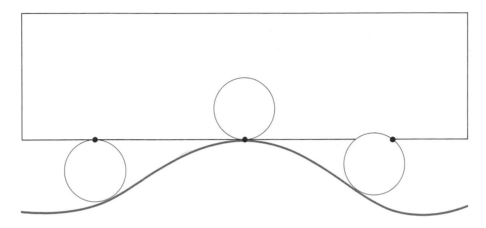

97. On what kind of curve will the car remain level?

three tracks lead into one horizontal track, three differently colored marbles will invariably enter the single track in the same order: the cycloid marble will lead, followed by the marble traveling on the circular arc and then by the one on the straight line.

The cycloid has other mechanical properties of interest. It is, as Galileo guessed, the strongest possible arch for a bridge, and for this reason many concrete viaducts have cycloidal arches. Cogwheels are often cut with cycloidal sides to reduce friction by providing a rolling contact as the gears mesh.

We have seen how a circle, rolled on a straight line, generates a cycloid. Stanley C. Ogilvy reverses this situation in one of his books by asking: Along what kind of curve can a circle be rolled so that a point on its circumference traces a straight line? To dramatize this question, imagine a railroad car with each wheel attached at its rim to the axle, as shown in Figure 97. How shall we curve a track so that when this curious car is rolled along the track it will remain level at all times and never bob up and down?

Answer

What kind of track will enable the car to travel without bobbing up and down? Figure 98 supplies the surprising answer: a series of semicircles! If a circle is rolled inside a circular arc, points on its circumference generate what are called hypocycloids.

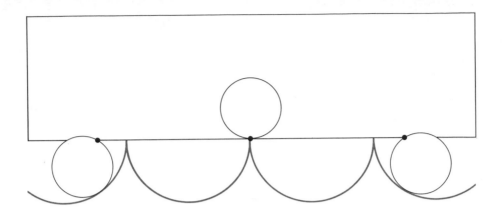

98. Solution to the car problem

When the radius of a semicircular track is twice that of the rolling circle, as it is here, the hypocycloid is a straight line.

References

A Treatise on the Cycloid. Richard Anthony Proctor. London: Longmans, Green and Co., 1878.

"Some Historical Notes on the Cycloid." E. A. Whitman. *The American Mathematical Monthly*, Vol. 50, No. 5; May, 1943. Pages 309–315.

A Book of Curves. E. H. Lockwood. Cambridge: Cambridge University Press, 1961.

"Brachistochrone, Tautochrone, Cycloid — Apple of Discord." J. P. Phillips. *The Mathematics Teacher*, Vol. 60, No. 5; May, 1967. Pages 506–508.

14. Mathematical Magic Tricks

MAGIC TRICKS that operate wholly or in part by mathematical principles fascinate a large segment of the conjuring fraternity. Dozens of such tricks are published every year in periodicals on magic or circulate from magician to magician, only occasionally finding their way into mathematical circles. Royal V. Heath's *Mathemagic* (1933) was the first book in this hybrid field. My own *Mathematics, Magic and Mystery* (1956) was the second. In 1964 Scribner's brought out a third: *Mathematical Magic*, by William Simon, who is president of a New Jersey firm that makes brake linings and also one of the country's most knowledgeable card experts.

Most of the items in Simon's fine collection will be unfamiliar to devotees of recreational mathematics. An example is a bewildering mind-reading trick discussed in the author's chapter on mental magic. Invented by Robert Hummer, a magician now living in Havre de Grace, Maryland, this trick is not only an entertaining parlor stunt but also such a puzzling exercise in

logic that many magicians who regularly perform the trick are not sure themselves just why it works.

One of the best presentations is as follows. Three identical coffee cups are inverted in a row on a table. The positions (not the cups) are assumed to be one, two and three as seen by the spectators [*see Figure* 99]. The magician, standing across the room with his back to the table, asks that a spectator conceal a small object, say a matchbook, under any one of the cups. The spectator now scrambles the positions of the cups by exchanging them in pairs, calling out each time the positions of the two cups involved. In making these exchanges the cups are slid across the table, so that if the cup covers the object, the object slides along with the cup. For example, suppose the matchbook is placed under the middle cup. If the spectator switches the end cups, he calls out, "One and three." If he next switches the two cups on his left, he calls out, "One and two." As these cups are slid the matchbook is carried along with

99. Hummer's three-cup trick

its cup from position two to position one. The spectator continues to switch pairs of cups as long as he wishes. The magician then turns around and immediately lifts the cup covering the matchbook. The trick can be repeated many times. Since the performer is never told which cup the object was placed under initially, how does he guess correctly?

The method is simple and subtle. Although the three cups are alike, it is impossible for them to be *exactly* alike. Inspect any three cups carefully and you are sure to find some tiny distinguishing feature — a small chip, a discoloration, and so on — on one of them. Before you turn your back note the position of this marked cup. After the matchbook has been placed under a cup explain the switching procedure to the spectator, then ask him to make a practice switch by exchanging the two *empty* cups. Caution him not to tell you the two positions, since that would give away the location of the matchbook. This practice switch seems to have no bearing whatever on the trick; in fact spectators usually forget it was even made. Actually it is the key to the trick, for a reason that I shall ask you to deduce.

As the spectator proceeds with his switch-

ing, calling out the positions of the cups each time, you must secretly keep track of one cup by using your left hand as a computer. Fingers one, two, and three represent positions one, two, and three. Start with the tip of your left thumb pressed against the finger tip that indicates the initial position of the marked cup. Of course, you must assume the marked cup is still in that position. Suppose at the start this cup is in the middle. You touch your second finger. If he calls one and two, move your thumb from the second to the first finger. If he next calls one and three, shift to the third finger. If he now calls one and two, you do not move your thumb: the position of the cup you are following is not involved in the exchange. When the spectator decides to stop, let us say your thumb touches your third finger.

Turn around and inspect the cups. If the marked cup is at position three, where your thumb says it should be, you know that this cup covers the matchbook. If the marked cup is not at position three, the object will be under the *unmarked* cup that is *not* at position three. (Can you explain why?)

Some magicians carry an artificial eye in their pocket to use in this trick. The performer uses the eye as the object placed under one of the cups; he can then en-courage the inference that the eye is some-how able to provide him with a clue to its own whereabouts. The eye also furnishes an excuse for amusing chatter. The ma-gician can say: "Yes, I see the evil eye staring at me from inside *this* cup. . . ."

Harry Lorayne, a New York City mne-monics expert (well known in entertain-ment circles for his sensational memory act), devised the following variation in which three objects are used instead of cups, and the magician is able to name the thought-of object without turning around. Three different objects—say a coin, a match-book, and a finger ring—are placed in a row and someone is asked to think of one of them. He must also be able to recall the order of the objects, or else he should jot it down for future reference. The performer turns his back and calls for a practice switch with the two objects the spectator did *not* think of. In this instance the spectator does not say what switch he has made. The trick then continues as with the three cups, the spectator making exchanges and calling out positions. When he finishes, the per-former asks if the objects are by any chance back in their original order. If not, the spec-tator makes the one or two additional switches needed to restore this order. These exchanges are called out as before. The performer seems to have no relevant information—the objects have merely been switched around and brought back to their initial state—yet he can name the thought-of object *without turning around*.

The method: Memorize the initial or-der. Pick any object and follow it with your thumb. You will not know, of course, whether or not that object remained in its original position after the practice ex-change. Nonetheless, after the original order has been restored, if your thumb indicates that the object you are following is back in its former position, you know that it is the chosen object. Otherwise the selected object is the one at the position represented neither by where your thumb started nor where it ended. Again can you explain why?

Before writing this chapter I got in touch with Robert Hummer and obtained his per-mission to describe another of his curious mind-reading tricks, here explained in print for the first time. The trick uses a card-board circle attached to a sheet of cardboard by a paper fastener through the center. On the rim of the circle, in any order, are inscribed the values of the 26 red playing cards. Outside the circle, on the backing sheet, are the 26 letters of the alphabet. They too may be in any order, but Hummer arranges them as shown in Figure 100 on the next page so that the 10 letters at the top spell "Think a word."

A spectator is asked to think of any word, preferably a short word of four or five let-ters. He also thinks of any red card. While the magician turns his back, the spectator rotates the wheel until his chosen card indicates the first letter of his word. The magician turns around, glances quickly at the dial, then turns his back again while the spectator moves the wheel so that his card points to the second letter of his word.

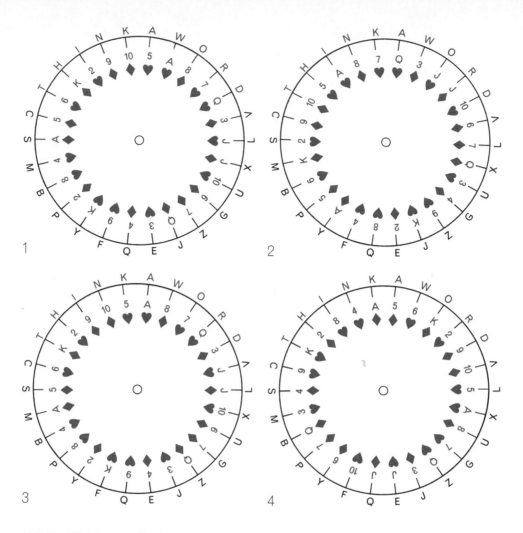

100. The "Think a word" trick

Again the magician glances at the dial. Obviously he does not know the spectator's card, so the dial would seem to give him no useful information. This procedure is repeated until the entire word is spelled. The magician, after appearing to concentrate for a moment, names both the word and the card.

A mathematician working with combinatorial mathematics, or a person skilled in cryptography, should have little difficulty devising a method for performing the wheel trick. For others I give it as a puzzle. The four positions of the dial in the illustration are typical of what the performer may see during the spelling of a four-letter word.

What word is being spelled there? It is not hard to find the word by the laborious procedure of testing each of the 26 red cards, but the problem is to devise a method that will enable the performer to name the word in a few seconds after seeing the dial's final position.

One of the best of many mathematical tricks invented by Jack Yates, a British magician, is his 12-penny trick, explained by Simon in a chapter on tricks with ordinary objects. The 12 pennies are arranged heads up in a circle to indicate the 12 hours on a clock. The penny at 12 o'clock is marked with a key as shown in Figure 101. While the performer's back is turned someone is asked to turn over any six coins. The magician, keeping his back turned, now gives directions for six more reversals. These are likely to involve some of the pennies reversed by the spectator; that is, some pennies turned tails up by the first six reversals may get turned back to heads by the second six reversals.

"How many heads are now showing?" the magician asks.

Suppose he is told: "There are two heads." Obviously the performer has no way of knowing which coins are heads and which are tails. Yet he is able to give directions for dividing the coins into two sets of six coins each so that the number of heads (and tails) in each set is the same. In this case each set would have one head and five tails.

Surprisingly, the performer does not need to be told the number of heads showing, but his asking for this information throws

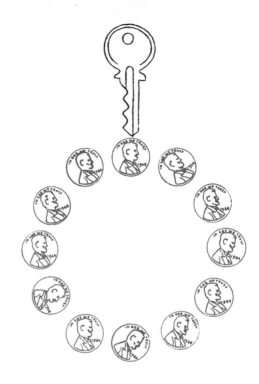

101. Yates's 12-penny trick

spectators off the track of a solution of the trick. When he directs the reversal of six coins, he may pick any six he wishes, but he must remember their numbers. For example, he may ask for the reversal of coins one, four, five, eight, nine and ten. To divide the coins properly into the two final sets he asks that the following six coins be slid to one side: 2, 3, 6, 7, 11, and 12. These are merely the six that are not in the previous set. (In set theory they are said to form the "complement" of the previous set.) To disguise the nature of this second set the performer directs their removal in pairs indicated by the hour of day. Instead of saying coins two and three, for instance, he says: "Please slide to one side the coins that mark ten minutes past three."

The principles of set theory exploited in this trick are the basis for numerous card tricks. The following, contributed by the British magician Norman MacCleod to a magic magazine in the United States, *The New Phoenix* (No. 328, August, 1955), is one of the best. While someone deals the deck into four bridge hands the performer writes on a slip of paper: "There will be 22 face-up cards." This prediction is folded and placed aside. A spectator takes two of the piles, the magician takes the other two.

"I have selected a number from one to ten," says the performer. "I shall turn that number of cards face up in each of my piles." He proceeds to turn some cards face up but without letting anyone see how many.

The spectator is asked to do the same with his two piles: choose a number from one to ten and reverse that number of cards in each pile. The four piles are assembled, the deck spread and the face-up cards counted. There are 22. The prediction is unfolded and found to be correct.

To perform this trick you must cheat a bit. Any even number between 13 and 39 can be written in your prediction. This number, minus 13, tells you the total number of cards to leave *face down* in your two packets. In this case 22 minus 13 is 9, so you reverse, say, all but 5 cards in one pile and all but 4 in the other. Put your two piles together and one of the spectator's piles on top. Hold this large packet in your left hand and ask the spectator to cut his remaining pile into two parts. While attention is focused on the cutting casually turn over your left hand, thus secretly reversing all its cards. This large pile is sandwiched between the two halves of the cut pile.

All the cards are now together again and presumably no one knows how many of them are face up. Do you see why there must be 22? The procedure reverses 13 cards in the spectator's two piles for the same reason that Yates's coin trick works. The 9 cards you left face down are now face up, making 22 in all. The trick can be repeated using other even numbers in the prediction. Odd numbers should be avoided because the procedure, if it is done legitimately, could not produce an odd number of face-up cards.

The magic linking and unlinking of rings can, if one stretches the term a bit, be regarded as topological effects. I have space for one quick trick invented by William Bowman, a Seattle magician, and described in Simon's chapter on topological magic. Attach two paper clips to a one-dollar bill in the manner shown in Figure 102. If the bill is held at the ends and pulled flat, the clips pop off the bill *linked together*. (The linking is puzzling enough, but why do the clips hop from the bill with such force?) Simon has a story of young love to provide patter for all this, but I prefer to have the spectator hold the bill so that the clips point down. When the bill is pulled flat, the clips drop to the floor. Bet even money they will fall within one inch of each other. Of course you can't lose.

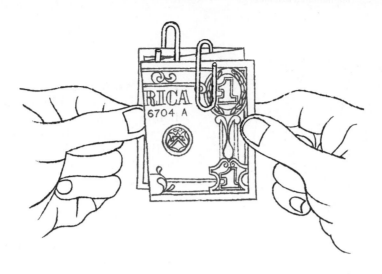

102. Bowman's bill trick

Answers

In the illustration for the "Think a word" trick the four-letter word being spelled is "love" and the thought-of card is the jack of hearts. To determine the word the magician uses a gimmick: a cylinder of five disks that rotate around a pin as shown in Figure 103. The 26 red cards are in the same order around the rim of the first disk as they are on the spelling wheel, and the 26 letters on each of the other disks are in the same order as those that surround the spelling wheel.

When the first letter of the word is spelled, the magician glances at the wheel and notes the letter opposite any card whatever, say the ace of hearts. As soon as his back is turned he rotates the second disk of his gimmick until this letter touches the ace of hearts. On his second glance at the wheel he notes the new letter opposite the ace of hearts. When his back is turned again, he adjusts the third disk accordingly. Similarly for the remaining two letters. In other words, the performer himself picks a card and uses it to spell four letters. He adjusts his dials so that his card and these four letters are in line. Then he turns the entire cylinder until he sees a four-letter word. It will be the word the spectator spelled. There is, of course, a chance that more than one word will turn up, but the odds are heavily against it. If it should happen, the magician simply makes more than one guess.

The gimmick can be made small enough to keep concealed in one hand. A similar gimmick can be made by mounting four concentric circles of graduated size on a

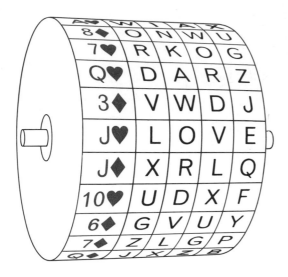

If the word is long, one is often able to spot the only possible combination of letters before the spelling is completed and so guess the word. In such cases a final look at the wheel will verify the guess, then the performer can proceed to name the word without turning his back again.

Some magicians omit the card symbols entirely from the gimmick. This has no effect on their ability to guess the word, and if someone asks them if they also know the selected card, they can answer, in complete honesty, that they haven't the slightest idea what is is!

103. Gimmick for the "Think a word" trick

square of cardboard. This can be kept in the performer's inside coat pocket, to be pulled out and secretly adjusted each time his back is turned. By adding more disks to the cylinder, or circles to the cardboard, one can do the trick with longer words.

References

Mathemagic. Royal Vale Heath. New York: Dover, 1953.

Mathematics, Magic and Mystery. Martin Gardner. New York: Dover, 1956.

Math Miracles. Wallace Lee. Durham, North Carolina: Privately printed, 1950. Revised edition, 1960.

Mathematical Magic. William Simon. New York: Charles Scribner's Sons, 1964.

15. Word Play

Was I clever enough? Was I charming? Did I make at least one good pun?

John Updike, *Thoughts while Driving Home*

WORD PLAY—puns, anagrams, palindromes and so on—is not discussed in any mathematics book, yet it has about it a quasi-mathematical air. Letters are symbols that combine according to rules to form words; words are symbols that combine according to rules to form sentences. Perhaps this combinatorial aspect is the reason so many mathematicians are addicted to language play.

The impulse to pun can persist even in the face of imminent death. On March 22, 1963, a murderer named Frederick Charles Wood was executed at Sing Sing. According to newspaper accounts, just before seating himself in the electric chair Wood said to those present: "I have a speech to make on an educational project. You will see the effect of electricity on Wood."

Less grim was the *New York Times* report a month later (April 28) that a gnu in the Chessington Zoo in England had bitten a zoo keeper. Odd, said the keeper, "most gnus are good gnus." I also find in my files an Associated Press dispatch from Des Moines, dated October 11, 1960, reporting that a perfume-dispensing machine in the women's lounge of a local hotel had failed to work. The management had hung a sign on it that read "Out of order." An unidentified patron, using lipstick, had crossed out the first "r" of "order."

The last is not strictly speaking a pun but rather a crude example of what word puzzlists call a deletion: the changing of one word into another by the removal of a letter. An amusing deletion story is told about Lord Kelvin, the British mathematician and physicist. He once put a sign on the door of a lecture hall stating that he would be unable to "meet my classes today." A student beheaded the word "classes" by crossing out the "c." Next day, eager to observe the professor's re-action, the students found that he had one-upped them by performing a second beheading.

The following is unusual: "Show this bold Prussian that praises slaughter, slaughter brings rout. Teach this slaughter lover his fall nears." If each word is be-headed, two entirely new sentences result. It is startling to learn that "startling" can be changed into eight other familiar words by successive deletions (from different places) of single letters. George Canning, an early-nineteenth-century British states-man, wrote the following verse about a word that is subject to "curtailment," that is, a word that becomes a different word when its last letter is removed. Can you identify the word?

> A word there is of plural number,
> Foe to ease and tranquil slumber;
> Any other word you take
> And add an "s" will plural make.
> But if you add an "s" to this,
> So strange the metamorphosis,
> Plural is plural now no more,
> And sweet what bitter was before.

Both decapitation and curtailment are involved in the following old riddle:

> From a number that's odd,
>   cut off the head,
> It then will even be;
> Its tail I pray now take away,
> Your mother then you'll see.

It would be interesting to know how many technical books of recent years have messages concealed in the text by playful authors. One finds out about them by acci-dent. Who would have guessed, for example, that *Transport Phenomena*, a 780-page textbook by R. Byron Bird, Warren E. Stewart and Edwin N. Lightfoot (published by John Wiley and Sons in 1960), had "On Wisconsin" hidden on page 712? (It is spelled by the first letters of each para-graph.) Or that the first letters of each sentence in the preface spell "This book is dedicated to O. A. Hougen"?

Sometimes word play enters a technical book fortuitously. Recently I had occasion to look up something in Rudolf Carnap's great work on semantics, *Meaning and Necessity*. On page 63 I came across a stretch of text in which the views of Black are sharply contrasted with those of White. Surely these were hypothetical individuals introduced to clarify an obscure point. No, on closer inspection they turned out to be the well-known philosophers Max Black and Morton White!

A classic instance of accidental word play is provided by the first (1819) edition of William Whewell's *Elementary Treatise on Mechanics*. On page 44 the text can be arranged in the following form:

> There is no force, however great,
> Can stretch a cord, however fine,
> Into a horizontal line,
> Which is accurately straight.

The buried poem was discovered by Adam Sedgwick, a Cambridge geologist, who recited it in an afterdinner speech. Whewell was not amused. He removed the poem by altering the lines in the book's next printing. Whewell actually published two books of serious poetry, but this unintended doggerel is the only "poem" by him that anyone now remembers.

If you keep your ears tuned, accidental meters turn up more often than you would expect. Max Beerbohm's eye caught the unintended beat in the following lines on the copyright page of the first English edition of one of his books:

> London: John Lane, The Bodley Head
> New York: Charles Scribner's Sons

Beerbohm completed the quatrain by writing

> This plain announcement, nicely read,
> Iambically runs.

"Quintessential light verse," wrote John Updike, commenting recently on the above lines, "a twitting of the starkest prose into perfect form, a marriage of earth with light, and quite magical. Indeed, were I a high priest of literature, I would have this quatrain made into an amulet and wear it about my neck, for luck."

The spoonerism, in which parts of two words (usually first syllables) are switched, continues to flourish as a popular form of wit. In 1960 Adlai Stevenson was campaigning in St. Paul, Minnesota, when the clergyman Norman Vincent Peale made some unfortunate political remarks. Stevenson told the press that he found St. Paul appealing and Peale appalling, surely one of the finest of all topical spoonerisms. In 1962, shortly after Rembrandt's painting "Aristotle Contemplating the Bust of Homer" had been bought by New York's Metropolitan Museum of Art for $2,300,000, it seems that Aristotle Onassis, the Greek shipping magnate, was shown Buster Keaton's house by a real estate agent. It was widely reported that a photograph in a Los Angeles newspaper was captioned "Aristotle Contemplating the Home of Buster," although I cannot vouch for it.

Ogden Nash's verse abounds in splendid spoonerisms:

> . . . I am a conscientious man,
>   when I throw
> rocks at sea birds
>   I leave no tern unstoned,
> I am a meticulous man
>   and when I portray
> baboons I leave no stern untoned.

No discussion of word play should fail to mention James Joyce. *Finnegans Wake* has, by a conservative estimate, 200 verbal plays per page, or more than 125,000 all together. The mathematical section of this book, pages 284 to 308 of the edition published by the Viking Press in 1947, contains hundreds of familiar mathematical terms, scrambled with metaphysics and sex. (The geometric diagram on page 293 is discussed mainly as a sex symbol.) The first footnote, "Dideney, Dadeney, Dudeney," refers to Henry Ernest Dudeney, the great English puzzle expert of Joyce's day. On page 302 "Smith-Jones-Orbison?" alludes to one of Dudeney's most popular puzzles, a logic problem involving three men named Smith, Jones and Robinson. Another of Dudeney's puzzles turns up in a footnote on page 299: "Pure chingchong idiotism with any way words all in one soluble. Gee each owe tea eye smells fish. That's U."

The puzzle: If you pronounce "gh" as in "tough," "o" as in "women" and "ti" as in "emotion," how do you pronounce "ghoti"? Was Joyce, in this footnote, speaking of the book itself and calling his reader a poor fish for biting the hook?

There are many references in *Finnegans Wake* to Lewis Carroll, who, as everyone knows, was a mathematician. In the mathematics section we read (page 294): "One of the most murmurable loose carollaries ever Ellis threw his cookingclass." (I scarcely need to point out that the last phrase puns on Alice *Through the Looking-Glass*.)

The following excerpt is from page 284: ". . . palls pell inhis heventh glike noughty times ∞, find, if you are not literally cooefficient, how minney combinaisies and permutandies can be played on the international surd! pthwndxrclzp!, hids cubid rute being extructed, taking anan illitterettes, ififif at a tom. Answers, (for teasers only)."

A partial explication: Pell was a mathematician for whom the Pellian equation was named, a number theorem often mentioned by Dudeney. "Heventh" is a compression of "seventh heaven." "Pthwndxrclzp" is one of the book's many thunderclaps. "Taking anan illitterettes, ififif at a tom" is, I suppose, "taking any letters, fifty at a

time." "For teasers only; is a play on "for teachers only."

The pangram, an ancient form of word play, is an attempt to get the maximum number of different letters into a sentence of minimum length. The English mathematician Augustus De Morgan tells (in his *A Budget of Paradoxes*) of unsuccessful labors to write an intelligible sentence using every letter once and only once. "Pack my box with five dozen liquor jugs" gets all 26 letters into a 32-letter sentence, and "Waltz, nymph, for quick jigs vex Bud" cuts it to 28. Dmitri Borgmann of Oak Park, Illinois, the country's leading authority on word play, has devised a number of 26-letter pangrams, but all require explanation. His best is "Cwm, fjord-bank glyphs vext quiz." A "cwm" is a circular valley, "quiz" is an eighteenth century term for an eccentric, a "glyph" is a carved figure. Borgmann's sentence thus states that an eccentric person was annoyed by carved figures on the bank of a fjord in a circular valley. Can any reader supply a better 26-letter pangram?

Another old and challenging word curiosity is the palindrome, a sentence that is spelled the same backward and forward. Borgmann's collection, covering all major languages, runs to several thousand. In my opinion the finest English palindrome continues to be "A man, a plan, a canal — Panama!" It has recently been attributed to James Thurber, but it was composed many years ago by Leigh Mercer of London, one of the greatest living palindromists. An unpublished Mercer palindrome, which is also something of a tongue twister, is "Top step's pup's pet spot."

Another Mercer palindrome, remarkable for both its length and naturalness, is "Straw? No, too stupid a fad. I put soot on warts." J. A. Lindon of Weybridge, England, is another master palindromist who turns them out by the hundreds. Who would suspect a palindrome if, in a novel, he came on the following Lindon sentence: "Norma is as selfless as I am, Ron." Lindon has also composed a large number of palindromes in which words rather than letters are the units. For instance: "So patient a doctor to try to doctor a patient so" and "Amusing is that company of fond people bores people fond of company that is amusing."

Composing anagrams (a phrase or word formed by rearranging the letters of another) on the names of friends or prominent people was once a fashionable literary sport. De Morgan tells of a friend who composed 800 anagrams on "Augustus De Morgan" (sample: "O Gus! Tug a mean surd!"). Lewis Carroll proudly recorded in his diary for November 25, 1868, that he had sent to a newspaper an anagram "which I thought out lying awake the other night: William Ewart Gladstone: Wilt tear down *all* images? I heard of another afterwards, made on the same name: 'I, wise Mr. G., want to lead all'—which is well answered by 'Disraeli: I lead, Sir!'" When Grover Cleveland was president, someone turned his name into "Govern, clever lad!" Theodore Roosevelt anagrams to "Hero told to oversee" and Dwight D. Eisenhower to "Wow! He's right indeed!" During the 1936 elec-

tion, Borgmann also informs me, the letters of Franklin Delano Roosevelt's name were permuted—by a Republican, no doubt—to "Vote for Landon ere all sink!" It was said during this campaign that the Republicans avoided picking Styles Bridges, at that time governor of New Hampshire, for Landon's running mate for fear the Democrats would go about chanting "Landon-Bridges falling down."

What can readers do with the full names of the two candidates for the 1964 election: Lyndon Baines Johnson and Barry Morris Goldwater?

For less ambitious readers Figure 104 on page 148 presents eight remarkable English words, the missing letters to be supplied. All letters omitted from the first word are consonants. The second word contains the first five letters of the alphabet in order. The third word can be typed by using only the top row of keys on a standard typewriter. (The letters of this row, left to right, are *QWERTYUIOP*.) The fourth and fifth words contain four letters in adjacent alphabetical order. The sixth word contains the five vowels in reverse order, the seventh the five vowels plus *Y* in the usual order. In the last word consonants and vowels alternate.

Addendum

The following letter appeared in the November 1964 issue of *Scientific American:*

Sirs:

Martin Gardner's department "Mathematical Games" is the first thing we look at when we

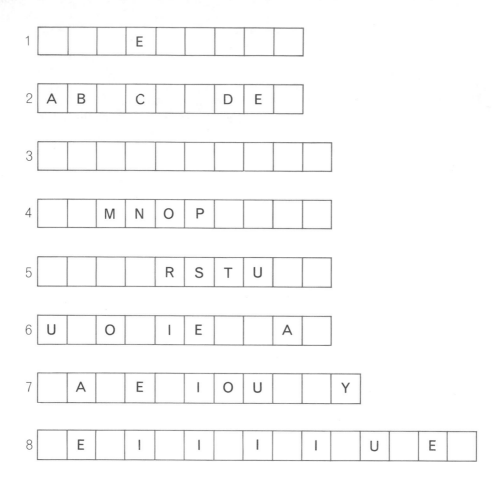

1. | | | | E | | | | |

2. | A | B | | C | | D | E | |

3. | | | | | | | | | |

4. | | | M | N | O | P | | | |

5. | | | | R | S | T | U | | |

6. | U | | O | | I | E | | A | |

7. | | A | | E | | I | O | U | | Y |

8. | | E | | I | | I | | I | | I | | U | | E | |

104. Eight curious words

pick up a copy of *Scientific American.* His September article on puns, palindromes, and other word games was quite entertaining and of particular interest to us.

In connection with the concealed message in the text *Transport Phenomena* by Bird, Stewart and Lightfoot, you might be interested to know that in the preface the first letters of the sentences actually spell "This book is dedicated to O. A. Hougen TTTM." The TTTM means "This terminates the message." Furthermore, in the forthcoming Spanish edition of our text *(Fenómenos de Transporte),* to be published by Editorial Reverté, the translator, Professor F. Mato Vázquez of the University of Salamanca, has obligingly translated our preface so that the hidden message is faithfully retained as *"Este libro está dedicado a O. A. Hougen,"* with no letters such as TTTM left over. In the postface we were faced with a problem, since "On Wisconsin" would have little meaning to Spanish-speaking readers and "w" does not occur in Spanish. Hence we have requested the translator to try to include the hidden message *"Adiós amigos"* instead.

You might also be interested to know that our colleague Professor Daizo Kunii (Department of Chemical Engineering, University of Tokyo) published a book entitled *Ryudo Kahyo* several years ago. The first characters of the paragraphs in his preface spell out, in Japanese, a dedication to his wife.

R. BYRON BIRD
WARREN E. STEWART
EDWIN N. LIGHTFOOT

Department of Chemical Engineering
University of Wisconsin
Madison, Wisconsin

Answers

The first of the two rhymed riddles is answered by the word "caress," the second by the word "seven." The six words in the closing quiz are "strengths," "absconder," "typewriter," "gymnoplast," "understudy," "unoriental," "facetiously" and "verisimilitudes."

A number of *Scientific American* readers responded to the request for pangrams. Walter G. Leight of the Franklin Institute's Center for Naval Analysis sent *Cozy sphinx waves quart jug of bad milk* (32 letters), *Blowsy red vixens fight a quick jump* (30) and *Quick jigs for waltz vex bad nymph* (28). The last is an improvement over the similar pangram given on page 146 because it eliminates the name "Bud." Proper nouns, abbreviations, initials and so on are considered blots on pangrams.

John G. Fletcher of Pleasanton, California, sent the best 26-letter pangram, which he says is due to the mathematician Claude E. Shannon: *Squdgy fez, blank jimp crwth vox!* A crwth is a stringed instrument of Welsh origin. "Jimp" is a Scottish word for "thin," "slender," "delicate." ("I see thee dancing on the green, thy waist sae jimp,/ Thy limbs sae clean," wrote Robert Burns.) The sentence is spoken by a man of the Near East to his short, squat fez as he pulls it down over his ears to blank out the thin, delicate voice (notes) of a crwth being played nearby. Vic Reid, Jr., of New York City reports that while Caesar's legions were encamped one night by a northern lake, they were approached by 15 mermaids who tried vainly to persuade the men to dance with them on the water. A war correspondent cabled 26 letters to his Roman editor: *XV quick nymphs beg fjord waltz.*

Several readers called attention to other answers to the quiz about eight curious words. *Absconded* can, of course, be substituted for *absconder*. Dmitri Borgmann writes that in addition to *typewriter* the following ten-letter words can also be typed on the top row of letter keys: *proprietor*, *pepperwort*, *pepperroot* and *protopteri*. Others (from W. H. Shepherd of Manchester, England): *perpetuity*, *repertoire*, *perruquier*, *pewterwort* and *pirouetter*. Borgmann goes on to say that *gymnopedia*, *limnophile* and *somnopathy* are other ten-letter words with *mnop* in the same spot as in *gymnoplast* (although he prefers the sentence *I am no prude*), and *pareciously* and *materiously* are alternates for *facetiously* in having the vowels and *y* in alphabetical order.

Stuart G. Schaeffer found another, more timely solution to the "cares — caress" riddle, which he expressed in what he calls "shaggy doggerel":

> *A century and more ago*
> *Clairvoyant Englishmen did know*
> *That in the twentieth century*
> *Tranquillity would shattered be,*
> *And so suggested bitter noise*
> *Be changed to sweet and silent joys*
> *By adding modest and conceitless*
> *"S" to make the Beatles beatless.*

The virtuosity of readers in finding anagrams on the full names of the two presidential candidates makes it impossible to do justice to the hundreds of ingenious anagrams received. Curiously each candidate's name involves a similar difficulty: taking care of the five *N*'s in Lyndon Baines Johnson and the five *R*'s in Barry Morris Goldwater. Dmitri Borgmann's best one for Johnson is *No ninny, he's on job, lads.* Essentially the same anagram was submitted by Arthur Schulman, James H. Cochrane, and Raphael M. Robinson. *Hands on only nine jobs* was independently devised by Mrs. H. A. Morss, Jr., and Mr. and Mrs. Bruce D. Hainsworth; virtually the same phrase also came from Mrs. E. M. Cutler and many others. The best anti-Johnson anagram is from Walter I. Cole, Jr.: *None sin? Sly hand*

on job. I should add that Cole also sent the following anti-Goldwater anagram: *My star error — a glib word.*

The best anagram favorable to Goldwater — *Smart, bold, grey warrior* — was submitted by David Rabby, who also balanced it with a favorable Johnson anagram. Most Goldwater anagrams stressed a fear that his policies would provoke war. *Morbid story — larger war* was discovered by both Mrs. Cutler and L. E. Card. Among 39 clever anagrams contributed by Mr. and Mrs. Gerald Dantzic are *Wary world's rarer bigot; Orders big "moral war" try!* Other anagrams of similar import: *Sorry brew, Mr. Gladiator!* (Mrs. Coburn A. Buxton), *Bald, raw, gory terrorism* (Arthur Schulman), *Sly orator bred grim war* (James H. Cochrane), *Grab rest, moldy warrior* (Alan Wachtel, Phil Leslie). John de Cuevas sent *A great world! By mirrors?* Mr. and Mrs. Bruce D. Hainsworth: *Red Star big moral worry.* Raphael Robinson, a well-known mathematician, imagined the following message signed with Barry's first initial: *Glory! I storm rearward. B.* To which Robinson added the following prayer for a Goldwaterloo: *Lord, bar grim worst year!*

References

Handy-Book of Literary Curiosities. William S. Walsh. Philadelphia: J. B. Lippincott Company, 1892.

Playing with Words. Joseph T. Shipley. New York: Prentice-Hall, 1960.

Oddities and Curiosities of Words and Litera-

ture. C. C. Bombaugh. New York: Dover, 1961.

Language on Vacation. Dmitri A. Borgmann. New York: Charles Scribner's Sons, 1965.

Games for Insomniacs. John G. Fuller. New York: Doubleday, 1966.

Beyond Language. Dmitri A. Borgmann. New York: Charles Scribner's Sons, 1967.

300 Best Word Puzzles. Henry Ernest Dudeney. 1925. (Reprint. New York: Charles Scribner's Sons, 1968.)

16. The Pythagorean Theorem

Such a theorem as "the square of the hypotenuse of a right-angled triangle is equal to the sum of the squares of the sides" is as dazzlingly beautiful now as it was in the day when Pythagoras first discovered it, and celebrated its advent, it is said, by sacrificing a hecatomb of oxen—a method of doing honor to Science that has always seemed to me slightly exaggerated and uncalled-for. One can imagine oneself, even in these degenerate days, marking the epoch of some brilliant scientific discovery by inviting a convivial friend or two, to join one in a beefsteak and a bottle of wine. But a hecatomb of oxen! It would produce a quite inconvenient supply of beef.

Lewis Carroll, *A New Theory of Parallels*

THE FIRST CHAPTER of Arthur Schopenhauer's great philosophical work *The World as Will and Idea* contains a harsh attack on Euclid's method of proving propositions, and on the famous forty-seventh proposition in particular. This is the familiar theorem, usually called the Pythagorean theorem, that states that the square on the hypotenuse of a right triangle has an area equal to the combined areas of the squares on the other two sides. It is, of course, one of the oldest and most indispensable theorems in the whole of mathematics.

Euclid's proof, as many readers will recall from high school geometry textbooks, is rather complicated. Construction lines are drawn here and there, says Schopenhauer, for no apparent reason; then we are dragged through a long chain of deductive steps until suddenly the proof snaps shut on us like a mousetrap. We are compelled to admit that the conclusion is true, but we feel somehow cheated. We do not "see" its truth. According to Schopenhauer we are like a doctor who knows both a disease and its cure but has no understanding of why the

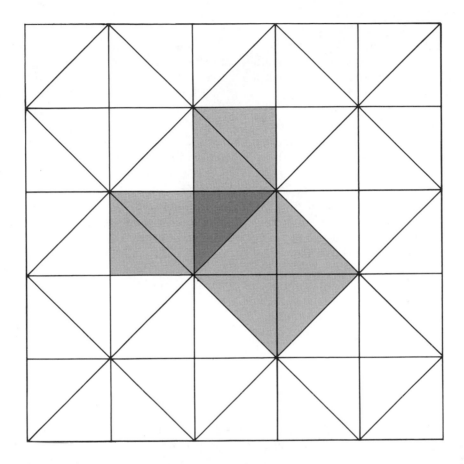

105. *Ancient Greek proof of Pythagorean theorem for the isosceles right triangle*

cure works. The proof is a "brilliant piece of perversity." It sneaks its truth in by a back door instead of giving it to us forthrightly, as a direct intuition of spatial relations.

A much better understanding of the theorem is obtained, Schopenhauer continues, by contemplating a diagram such as the one in Figure 105. We see at once that the squares on the two legs of the shaded triangle are composed of four congruent triangles that fit together to form the square on the hypotenuse. Essentially the same diagram is used by Socrates (in Plato's *Meno*) to convince a slave boy of the truth of a theorem. How foolish, Schopenhauer says, to toil over Euclid's rough terrain when we can get there directly by such a "bright, firm road."

Schopenhauer's arguments are naïve: the proof he recommends concerns only a special case, the isosceles right triangle, and does not prove the theorem at all. Nevertheless, there is something to be said for the pedagogic value of simple proofs that give a maximum of intuitive insight. Consider the figure at left in Figure 106. Clearly *any* type of right triangle can be duplicated four times and arranged in this pattern. The tilted white square in the center—the square on the hypotenuse—has an area equal to that of the large square minus the combined areas of the four shaded triangles. Now we rearrange the four triangles inside

the same large square in the manner shown in the figure at right in the illustration. The two white squares are the squares on the two legs. Since their combined area also is that of the large square minus the four triangles, we know it must equal the area of the tilted white square in the figure at left in the illustration.

No one knows who first thought of this beautiful proof, but it may predate Pythagoras himself. The figure at the left in the illustration appears in the *Chou Pei*, a Chinese manuscript that goes back to the Han period (202 B.C. to A.D. 220) but is believed to contain much older mathematical material. Although the manuscript gives no actual proof, it does mention the right triangle with integral sides of 3, 4, and 5, and many scholars think that the figure played

106. A "look-see" proof of the theorem for any type of right triangle

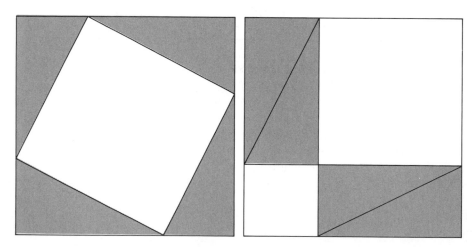

a part in a proof similar to the one just explained. Pythagoras, who lived about 500 B.C., is believed to have proved the theorem (legend has it that he sacrificed one hundred oxen when he first discovered the theorem), but no actual proof by him has survived. Recent research has disclosed that the ancient Babylonians, more than a thousand years before the time of Pythagoras, knew the theorem as well as many different kinds of right triangle with integral sides. There is *no* evidence that the Egyptians knew either the theorem or the 3, 4, 5 triangle. The myth that they did goes back to 1900, when Moritz Cantor, a German historian of mathematics, knowing that Egyptian temple builders used ropes in laying foundations, suggested that *perhaps* they obtained accurate right angles by using marked ropes that could be stretched around stakes to form a 3, 4, 5 triangle. Perhaps they did, but there is not a single known document to support this guess.

A delightful, dynamic proof of the theorem, devised by a New York mathematician, Hermann Baravalle, was published in 1945. Its five steps are shown in Figure 107. Only the fourth step calls for comment. If a parallelogram is altered by a shearing motion that preserves its base and altitude, its area remains constant.

I know of no more intuitively satisfying proofs of the theorem than these, but by applying some elementary algebra still simpler proofs are possible. Surely the simplest is obtained by resting the triangle on its hypotenuse, as shown in Figure 108,

then dropping a vertical line from the top corner. The small shaded right triangle is similar to the large triangle ABC because both have the angle A in common. Similar triangles have sides in the same ratio, therefore $b : x = c : b$, or $b^2 = cx$. The small unshaded right triangle is similar to ABC (they have angle B in common), therefore $a : c - x = c : a$, or $a^2 = c^2 - cx$. We add the two equations

$$
\begin{aligned}
b^2 &= cx \\
a^2 &= c^2 - cx \\
\hline
a^2 + b^2 &= c^2
\end{aligned}
$$

and obtain the theorem.

Hundreds of ingenious ways to prove the theorem have been published. The second (1940) edition of *The Pythagorean Proposition,* by Elisha S. Loomis, gives 367 different proofs, neatly classified by types. Of special interest—it is the only contribution to mathematics every made by a president of the United States!—is an algebraic proof based on the construction shown in Figure 109. The proof first appeared in a Boston weekly called *The New England Journal of Education* on April 1, 1876, with a note by the editor saying it had been given to him by James A. Garfield, then a Republican congressman from Ohio. Garfield had hit on it, says the note, during "some mathematical amusements" with other congressmen, and "we think it something on which the members of both houses can unite without distinction of party." The basic right triangle is shown shaded. On its hypotenuse is drawn the right isosceles triangle CBE. Line AC is extended, then from point E a

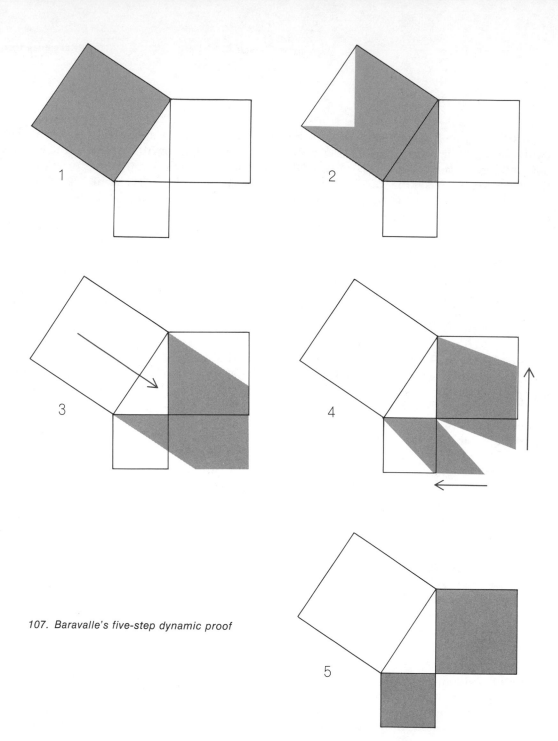

107. Baravalle's five-step dynamic proof

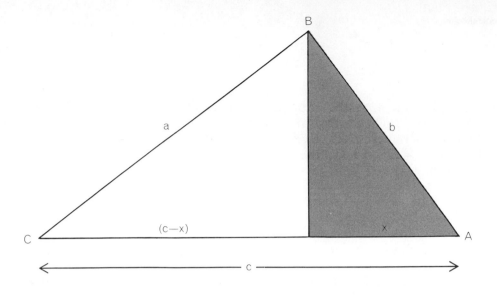

108. Simplest algebraic proof of the theorem

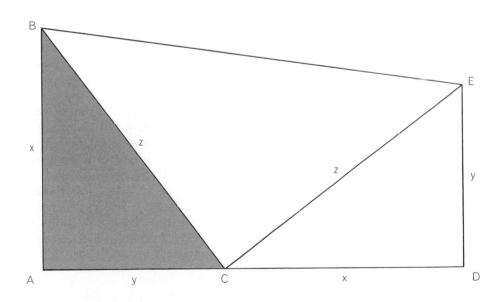

109. President Garfield's proof

perpendicular is drawn to the extended line, meeting it at *D*. The shaded triangle is congruent with triangle *DCE*, therefore *AB = DC* and *AC = DE*. I leave the proof as a puzzle for the reader.

The theorem can be generalized in scores of interesting ways. For instance, any figure can be drawn on the three sides — semicircles, hexagons, triangles and so on. As long as the three figures are similar, with corresponding sides on the triangle, the area of the figure on the hypotenuse must equal the sum of the areas of the other two. Pappus of Alexandria, a Greek geometer who lived about A.D. 300, proved a much more remarkable generalization. One starts with any triangle whatever [*ABC in Figure 110*]. On its legs one draws two parallelograms [*shown shaded*] of any size or shape. Sides of these two parallelograms are extended to meet at point *P*. We next draw a line through *P* and *C*, extending it downward until *QR* is equal to *PC*. If a parallelogram is drawn on the hypotenuse of the triangle, its sides equal to and parallel with *PR*, its area will be the sum of the areas of the other two parallelograms.

One proof is ridiculously easy. The shaded parallelogram at left in the illustration is equal in area to parallelogram *WPCA* (for the reason given in connection with Baravalle's proof) and also (for the same reason) equal to parallelogram *AQRX*. At right, the same argument shows that the shaded parallelogram has an area equal to parallelogram *QBYR*. Since the large parallelogram on the hypotenuse is made up of *AQRX* and *QBYR*, its area is the sum of the areas of the two shaded parallelograms. It is easy to see that the Pythagorean theorem is a special case of Pappus' theorem. It obtains when angle *C* is the right angle and the two shaded parallelograms are squares. In this special case the proof just outlined is essentially the same as Baravalle's proof.

The simplest right triangle with integral sides is the 3, 4, 5 triangle. Of course we can get an infinity of other "Pythagorean triples," as these three numbers are called, simply by multiplying each number by the same integer. If we multiply by 2, we get the Pythagorean triple 6, 8, 10. This is not very exciting, because a triangle with such sides is merely an enlarged version of the 3, 4, 5. Much more interesting are the Pythagorean triples that have no common factor, that is, that have integers that are "coprime." Such triples are called "primitive Pythagorean triples," which we abbreviate to PP triples. Obviously no two PP triangles will have the same shape.

Every Pythagorean triple, primitive or not, is an integral solution of the equation $x^2 + y^2 = z^2$. There is an infinite number of *primitive* solutions. (If the exponent of the three terms is any integer greater than 2, there are believed to be *no* integral solutions. This is Pierre de Fermat's famous "last theorem," not yet proved true.) The formula for finding primitive solutions goes back to the Greeks and probably back to ancient Babylonia:

$$x = a^2 - b^2$$
$$y = 2ab$$
$$z = a^2 + b^2$$

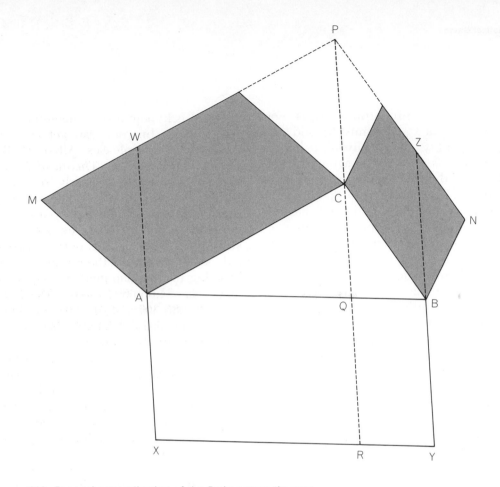

110. Pappus' generalization of the Pythagorean theorem

The letters x and y are the triangle's legs, z is the hypotenuse. Letters a and b stand for integers called "generators." They can be any pair of positive integers, with the restrictions that they be coprime (have no common divisor), of opposite parity (one even, one odd), and that a be greater than b. For example, if b is 1 and a is 2 (the smallest possible generators), we obtain the 3, 4, 5 triangle. Generators of 3 and 2

(for a and b respectively) give the next simplest PP triple: 5, 12, 13. In this way the formula generates all PP triples. There are 16 PP triangles with sides less than 100 and exactly 100 such Pythagorean triangles (including the primitives) if we count mirror images as being different.

The study of Pythagorean triples has long been a vigorous branch of recreational number theory, with a literature that has reached awesome proportions. It is not hard to prove that x and z must be odd and that y is "doubly even" (divisible by 4). Either x or y is sure to be a multiple of 3, and one of the three numbers must be a multiple of 5. Since the factors 3, 4, 5 occur somewhere in the triple, the product of all three numbers must be a multiple of 60. The area of a PP triangle must be a multiple of 6 and cannot be a perfect square.

Taking off from such simple properties, students of Pythagorean triples have set themselves an endless variety of bizarre problems. How many PP triangles have a certain integer as a leg? As a hypotenuse? Find PP triangles with a perimeter that is a square, or an area that equals the hypotenuse, or legs that differ by 1, or an area that contains each of the nine digits once and only once, and so on. It is difficult to invent a problem along such lines that has not been industriously worked on.

It is easy to prove, for instance, that only two Pythagorean triangles—6, 8, 10 and 5, 12, 13—have perimeters that equal their areas. Is there a PP triangle whose hypotenuse is a perfect square, and with legs such that their difference is also a square? Yes; the smallest such triangle is 119, 120, 169. Is there a PP triangle with a square hypotenuse and legs that *sum* to a square? Yes; but now the smallest answer is 4,565,486,-027,761, 1,061,652,293,520 and 4,687,298,-610,289. (This last problem was posed and solved by Fermat in 1643.) The PP triangle with sides 693, 1,924, 2,045 has an area of 666,666.

No isosceles right triangle can be Pytha-gorean (its hypotenuse is incommensurable with a leg), but one can get as close to isosceles as one pleases. Albert H. Beiler, in *Recreations in the Theory of Numbers,* gives a PP triangle so nearly isosceles that if the sides of one of its acute angles were extended 100 billion light-years, the divergence from a 45-degree angle would still be (as Beiler points out) an inconceiv-ably small fraction of the radius of a proton! One leg in this mammoth Pythagorean tri-angle is 21,669,693,148,613,788,330,547,-979,729,286,307,164,015,202,768,699,465,-346,081,691,992,338,845,992,696. The other leg is that number plus 1.

Some of the most challenging problems in the field concern PP triangles that have the same area. Fermat showed how to find a set of as many equiareal nonprimitive Pythagorean triangles as desired. Some 20 years ago William P. Whitlock, Jr., worked out a number of ingenious formulas for finding pairs of equiareal *primitive* Pytha-gorean triangles. So far, however, only one example has been found of *three* equiareal PP triangles: 1,380, 19,019, 19,069; 3,059, 8,580, 9,109; 4,485, 5,852, 7,373. Their common area is 13,123,110. (This triplet was discovered in 1945 by Charles L. Shedd of Arlington, Massachusetts.) Is there an-other triplet? Are there *four* equiareal PP triangles? No one knows.

You will want to leave these difficult questions to the experts. Here are four easy, although in some ways tricky, Pytha-gorean triangle problems, all answered in the answers section.

1. Which has the larger area, a triangle

with sides 5, 5, 6 or one with sides 5, 5, 8?

2. A 30, 40, 50 Pythagorean triangle has a perimeter of 120. Find two other Pythagorean triangles with the same perimeter.

3. What is the smallest number of matches needed to form simultaneously, on a plane, two different (noncongruent) Pythagorean triangles? The matches represent units of length and must not be broken or split in any way.

4. For all Pythagorean triangles the diameters of inscribed and circumscribed circles are integral. The diameter of the inscribed circle is obtained by adding the legs and then subtracting the hypotenuse (for example, the diameter of the circle inscribed in the 3, 4, 5 triangle is 2). Find a formula for the diameter of the circumscribed circle.

Answers

What is President James Garfield's proof of the Pythagorean theorem? Referring to the diagram on page 157, the area of the entire figure — trapezoid $ABED$ — is the product of its base, $x + y$, and half the sum of its sides, x and y. This can be written

$$\frac{(x + y)(x + y)}{2}.$$

The area of the trapezoid is also the sum of the areas of the three triangles. The largest triangle has an area of $z^2/2$, and each of the other two (congruent) triangles has an area of $xy/2$. We express the trapezoid's area as

$$\frac{z^2}{2} + \frac{2(xy)}{2}.$$

The two expressions for area are equal, so we have the equation

$$\frac{(x + y)(x + y)}{2} = \frac{z^2}{2} + \frac{2(xy)}{2},$$

which simplifies to

$$x^2 + y^2 = z^2.$$

I. Carl Romer, Jr., pointed out that Garfield's proof is essentially the same as the "look-see" proof in Figure 106. Garfield's figure is exactly one half of the figure on the left of the "look-see" illustration.

The four problems involving Pythagorean triangles are answered as follows:

1. Triangles 5, 5, 6 and 5, 5, 8 have equal areas because each can be split in half to make two 3, 4, 5 triangles.

2. The smallest Pythagorean triangles with the same perimeter are 30, 40, 50; 24, 45, 51, and 20, 48, 52. Each has a perimeter of 120. The three smallest primitive Pythagorean triangles with equal perimeters are 3,255, 5,032, 5,993; 7,055, 168, 7,057, and 119, 7,080, 7,081.

3. Two noncongruent Pythagorean triangles — 3, 4, 5 and 6, 8, 10 — can be formed simultaneously on the plane with as few as 27 matches [see Figure 111].

4. The diameter of a circle circumscribed about any right triangle is equal to the triangle's hypotenuse, as is evident from Figure 112.

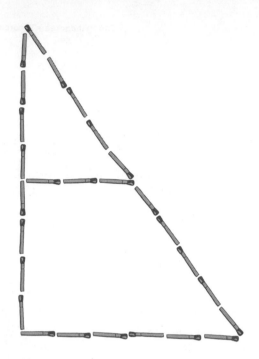

111. Answer to the match problem

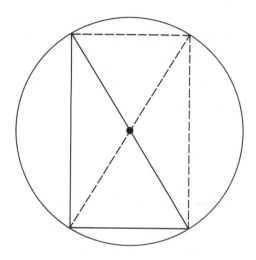

112. Circumscribing a right triangle

References

The Pythagorean Proposition. Elisha Scott Loomis. Privately printed, 1940. (Reprint. Washington, D.C.: National Council of Teachers of Mathematics, 1968.)

"Rational Right Triangles with Equal Areas." W. P. Whitlock, Jr. *Scripta Mathematica,* Vol. 9, No. 3; September, 1943. Pages 155–161.

"Rational Right Triangles with Equal Areas." W. P. Whitlock, Jr. *Scripta Mathematica,* Vol. 9, No. 4; December, 1943. Pages 265–268.

Pythagorean Triangles. Waclaw Sierpinski. New York: Academic Press, 1962.

"The Eternal Triangle." Albert H. Beiler. In his *Recreations in the Theory of Numbers.* New York: Dover Publications, 1964.

17.   Limits of Infinite Series

The Ball laughed. If you have never heard an india-rubber ball laugh you won't understand. It's the sort of quicker, quicker, quicker, softer, softer, softer chuckle of a bounce that it gives when it's settling down when you're tired of bouncing it.

E. Nesbit, *Nine Unlikely Tales*

FOR A MATHEMATICS STUDENT about to make the great leap from precalculus to calculus, no asset is more valuable than a firm, intuitive grasp of the concept of limit. The derivative and the definite integral, the fundamental tools of calculus, are both limits of infinite series. Every irrational number, such as pi, e, and the square root of 2, is the limit of an infinite series. Perhaps an approach to the concept by way of recreation will help to dispel some of the difficulties that caused so much metaphysical confusion in the early history of calculus and that are still stumbling blocks in the path of a student today.

It was Zeno of Elea, a Greek philosopher of the fifth century B.C., who first demon-strated, with a famous series of paradoxes, how easily one falls into logical traps in talking about an infinite series. How, Zeno asked, can a runner ever get from A to B? First he must go half the distance. Then he must go half the remaining distance, which brings him to the 3/4 point. But before completing the last quarter he must again go halfway, to the 7/8 point. In other words, he goes a distance equal to the sum of the following series:

$$1/2 + 1/4 + 1/8 + 1/16 + \ldots$$

The dots at the end mean that the series continues forever. How can a runner traverse an infinite series of lengths in a finite time? If you keep adding the terms of this series, you will never reach the goal of 1; you are always short by a distance equal to the last fraction added.

Now, there is a simple way to design an experiment so that in theory Zeno's contention is correct. Place a chess queen so that the center of its circular base rests on point A. The piece is to be pushed along a straight line to point B in the following way. First we push it a distance of 1/2, then pause until one second has elapsed. Then we push it a distance of 1/4 and again pause until the end of the second. We continue in this manner, beginning each push one second after the start of the previous push. At what time will the queen reach B? The answer is never. Suppose, however, we give the queen a constant velocity so that it covers half the distance in half a second, a quarter of the distance in a quarter of a second and so on. Both time and distance are now described by the same halving series. Both simultaneously converge—or "choke off," as mathematicians say—at the number 1. In one second, therefore, the queen reaches B.

What does a mathematician mean when he says that the "sum" of this halving series is 1? Clearly it is not a sum in the sense that one speaks of the sum of a finite series. There is no way to sum an infinite series in the usual sense of the word because there is no end to the terms that must be added. When a mathematician speaks of the sum— more precisely the limit—of an infinite series, he means a number that the value of the series approaches, as the number of its

terms increases without bound. By "approach" he means that the difference between the value of the series and its limit can be made *as small as one pleases*. Here we touch the heart of the matter. The value of an infinite series sometimes reaches its limit and sometimes goes *beyond* the limit. A simple example of the latter is obtained by changing alternate signs in the halving series to minus signs: $1/2 - 1/4 + 1/8 - 1/16 + \ldots$. The partial sums of this series are alternately more or less than its limit of .3333 . . . (which, incidentally, is a way of writing 1/3 as the limit of an infinite series of decimal fractions). The important point is that, in every case of an infinite series that chokes off, one can always find a partial sum that differs from the limit by an amount smaller than any fraction one cares to name.

Finding the limit of a converging series is often extremely difficult, but when the terms decrease in a geometric progression, as in the case of the halving series, there is a simple dodge every reader should know. First let x equal the entire series. Because each term is twice as large as the next, multiply each side of the equation by 2:

$$2x = 2(1/2 + 1/4 + 1/8 + 1/16 + \ldots)$$
$$2x = 1 + 1/2 + 1/4 + 1/8 + \ldots$$

The new series, beyond 1, is the same as the original series x. So

$$2x = 1 + x,$$

which reduces to $x = 1$.

Let us see how this applies to another of

Zeno's paradoxes: the race of Achilles and the tortoise. Assume that Achilles runs ten times as fast as the tortoise, and that the animal has a lead of 100 yards. After Achilles has gone 100 yards the tortoise has moved 10. After Achilles has run 10 yards the tortoise has moved 1. If Achilles takes the same length of time to run each segment of this series, he will never catch the tortoise, but if both move at uniform speed, he will. How far has Achilles gone by the time he overtakes the tortoise? The answer is the limit of the series $100 + 10 + 1 + .1 + .01 + .001 + \ldots$. Here we see at once that the sum is $111.111 \ldots$, or $111\frac{1}{9}$ yards. Suppose Achilles runs seven times as fast as the tortoise, which has the same head start of 100 yards. How far must Achilles go to catch the tortoise?

(We leave aside the question of whether modern mathematics does or does not refute Zeno. It all depends, of course, on what one means in this context by "refute." The interested reader can find no better introduction to the difficult literature on this subject than Bertrand Russell's brief discussion in Lecture 6 of *Our Knowledge of the External World* and his more advanced analysis on pages 336–354 of *Principles of Mathematics* (Second edition; New York: W. W. Norton and Company, 1938). Zeno's paradoxes raise questions about space, time and motion that are too deep to be answered frivolously, as they once were by Diogenes the Cynic: he stood up and walked from A to B.)

Bouncing-ball problems, found in many puzzle books, also yield readily to the trick just explained. Assume that an ideal ball is dropped from a height of one foot. It always bounces to 1/3 of its previous height. If each bounce takes a second, the ball will bounce forever, but since the time for each bounce also decreases by a converging series, the ball eventually stops bouncing even though it makes (in theory) an infinite number of bounces. The reader should have little difficulty determining how far this ideal ball travels before it comes to rest.

Geometric examples of series of this type are legion. If the largest square in Figure 113 has a side of 1 and the nesting continues indefinitely, what is the area of the infinite set of squares? Obviously it is 1 plus the halving series previously considered, or a total area of 2. Only a trifle more difficult is the following problem, presented in 1905 in a competition held annually in Hungary. A unit square is divided into nine equal squares, like a ticktacktoe board, and the center square is painted a color. The remaining eight squares are similarly divided and painted. If repetitions of this procedure continue indefinitely [*see Figure 114*], what is the limit of the painted area?

When a series does not converge, it is said to diverge. It is easy to see that $1 + 2 + 3 + 4 + 5 + \ldots$ does not choke off. Suppose, however, that each new term, in a series joined by plus signs, is *smaller* than the preceding one. Must such a series converge? It may be hard to believe at first, but the answer is no. Consider the series known as the harmonic series:

$$1 + 1/2 + 1/3 + 1/4 + 1/5 + \ldots$$

The terms get smaller and smaller; in fact,

113. *An infinite set of nested squares*

they approach zero as a limit. Nevertheless, the sum increases without bound! To prove this we have only to consider the terms in groups of two, four, eight, and so on, beginning with 1/3. The first group, $1/3 + 1/4$, sums to more than 1/2 because 1/3 is greater than 1/4, and a pair of fourths sums to 1/2. Similarly, the second group, $1/5 + 1/6 + 1/7$ + 1/8, is more than 1/2 because each term except the last exceeds 1/8, and a quadruple of eighths sums to 1/2. In the same way the third group, of eight terms, exceeds 1/2 because every term except the last (1/16) is greater than 1/16, and 8/16 is 1/2. Each succeeding group can thus be shown to exceed 1/2, and since the number of such groups is

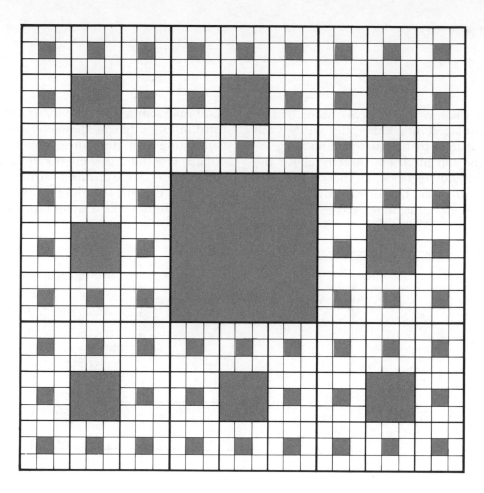

114. What is the limit of area for the colored portion?

unlimited the series must diverge. It does so, however, with infuriating slowness. The first 100 terms, for instance, total only a bit more than 5. To reach 100 requires more than 2^{143} terms, but less than 2^{144} terms. (I am indebted to Daniel Asimov for supplying these upper and lower bounds.) In 1968 John W. Wrench, Jr., calculated the exact number of terms at which the series has a partial sum exceeding 100. The number of terms is 15,092,688,622,113,788,323,693,-563,264,538,101,449,859,497.

The harmonic series is involved in an amusing problem that appeared in the *Pi Mu Epsilon Journal* for April, 1954, and more recently in *Puzzle-Math,* a book by

George Gamow and Marvin Stern. If one brick is placed on another, the greatest offset is obtained by having the center of gravity of the top brick fall directly above the end of the lower brick, as shown by arrow *A* in Figure 115. These two bricks, resting on a third, have maximum offset when their combined center of gravity is above the third brick's edge, as shown by arrow *B*. By continuing this procedure downward one obtains a column that curves in the manner shown. How large an offset can be obtained? Can it be the full length of a brick?

The unbelievable answer is that the offset

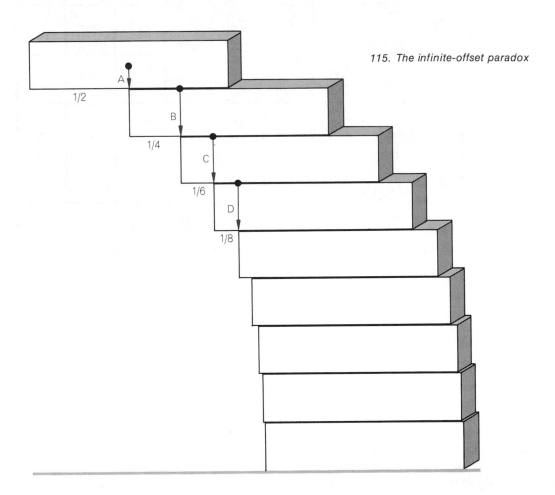

115. The infinite-offset paradox

can be as large as one wishes! The top brick projects half a brick's length. The second projects 1/4, the third 1/6 and so on down. With an unlimited supply of bricks the offset is the limit of

$$1/2 + 1/4 + 1/6 + 1/8 + \ . \ . \ .$$

This is simply the harmonic series with each term cut in half. Since the sum of the harmonic series can be made larger than any number we care to name, so can half its sum. In short, the series diverges, and therefore the offset can be increased without limit. As we have seen, such a series diverges so slowly that it would take a great many bricks to achieve even a small offset. With 52 playing cards, the first placed so that its end is flush with a table edge, the maximum overhang is a little more than 2¼ card lengths [*see Figure 116.*]. Readers may enjoy seeing if they can build an offset, using one deck, that exceeds two card lengths.

The harmonic series has many curious properties. If every term containing the digit 9 is crossed out, the remaining terms

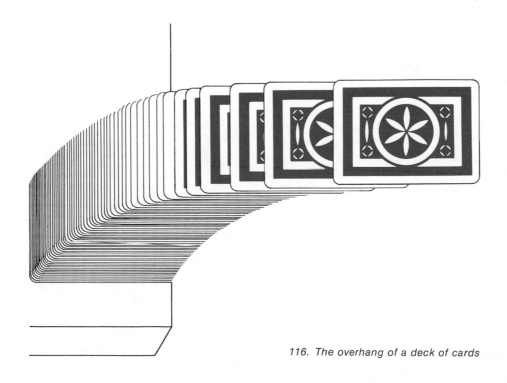

116. *The overhang of a deck of cards*

form a convergent series. If the denominator of each term is raised to the same power n, and n is greater than 1, the series converges. If every other sign, starting with the first, is changed to minus, the resulting series

$$1 - 1/2 + 1/3 - 1/4 + 1/5 - \ldots$$

chokes off on the natural logarithm of 2, a number slightly smaller than .7. Does the value of the series ever reach (after 1, of course) a number that is an integer? If there were a simple formula for expressing the value of the series for n terms, this might be easily answered, but there is no such formula. An ingenious odd-even argument, however, that goes back at least to 1915 (the details are given on page 48 of the *American Mathematical Monthly* for January, 1934) shows that the series never reaches an integral sum.

If all the terms of an infinite series are positive, it clearly does not matter how the terms are grouped or rearranged; the limit remains the same. But if there are negative terms, it sometimes makes a big difference. From the seventeenth century to the middle of the nineteenth, before laws of limits were carefully formulated, all sorts of disturbing paradoxes were produced by juggling the plus and minus terms of various infinite series. Luigi Guido Grandi, a mathematician at the University of Pisa, considered the simple oscillating series $1 - 1 + 1 - 1 + 1 - \ldots$. If one groups the terms $(1-1) + (1-1) + (1-1) + \ldots$, the limit is 0. If one groups them $1 - (1-1) - (1-1) - \ldots$, changing the signs within parentheses as

required, the sum is 1. This shows, Grandi said, how God could take a universe with parts that added up to nothing and then, by suitable rearranging, create something.

The correct limit for the original series, Grandi declared, is 1/2. He supported this by a parable. A father wills a precious stone to two sons with the proviso that every year the stone go from one to the other. If the value of the stone is 1, then its value to each son is the sum of $1 - 1 + 1 - 1 + \ldots$. Since the two brothers share the legacy equally, this value must be 1/2. Many distinguished mathematicians joined in the controversy over this series. Both Gottfried Wilhelm von Leibniz and Leonhard Euler agreed on the 1/2, although for somewhat different reasons. Today the series is recognized as divergent, so that no meaningful limit can be assigned to it.

An even worse instance is provided by the series $1 - 2 + 4 - 8 + 16 - \ldots$. Group it $1 + (-2 + 4) + (-8 + 16) + \ldots$ and you obtain the series $1 + 2 + 8 + 16 + \ldots$, which diverges to positive infinity. Group it $(1 - 2) + (4 - 8) + (16 - 32) + \ldots$ and you get the series $-1 - 4 - 16 - 64 - \ldots$, which diverges to infinity in the negative direction! The climax to all this infernal hubbub came in 1854 when Georg Friedrich Bernhard Riemann, the German mathematician now well known for his non-Euclidean geometry, proved a truly remarkable theorem. Whenever the limit of an infinite series can be changed by regrouping or rearranging the order of its terms, it is called *conditionally* convergent in contrast to an *absolutely* convergent series, which is unaffected by such scrambling. Conditionally convergent series always have negative terms, and they always diverge when all their terms have been made positive. Riemann showed that any conditionally convergent series (such as the one previously cited that chokes off on the natural logarithm of 2) can be suitably rearranged to give a limit that is any desired number whatever, rational or irrational, or even made to diverge to infinity in either direction.

Even an infinite series without negative terms, if it diverges, can cause serious trouble if one tries to handle it with rules that apply only to finite and converging series. For example, let x be the infinite, positive sum of $1 + 2 + 4 + 8 + 16 + \ldots$. Then $2x$ must equal $2 + 4 + 8 + 16 + \ldots$. This new series is merely the old series minus 1. Therefore $2x = x - 1$, which reduces to $x = -1$. Thus we seem to have proved that -1 is infinite and positive. One can sympathize with the Norwegian mathematician Niels Henrik Abel, who wrote in 1828: "The divergent series are the invention of the devil, and it is a shame to base on them any demonstration whatever."

Addendum

S. W. Golomb was the first of several mathematicians to point out that I was not quite accurate in saying that a divergent series could not be given a meaningful sum. "After convincing our undergraduates that divergent series are the invention of the devil," Golomb wrote, "we let them learn in graduate school that these series can be 'summed' after all, if one is sufficiently careful to define new kinds of summation rules (*e.g.*, Cesàro summation, Abel summation, etc.)." Golomb went on to say that G. H. Hardy's *Divergent Series* (New York: Oxford Press, 1949) was a remarkable book in which such summation techniques are explained. The series $1 - 1 + 1 - 1 + 1 \ldots$, for example, has both a Cesàro sum and an Abel sum of 1/2, as Leibniz and Euler maintained. The reader is referred to Hardy's posthumous book for a fascinating survey of the field.

Answers

If Achilles runs seven times as fast as the tortoise, which has a head start of 100 yards, the total distance Achilles travels, before overtaking the tortoise, is the limit of the series

$$100 + \frac{100}{7} + \frac{100}{7\cdot7} + \frac{100}{7\cdot7\cdot7} + \ldots$$

Each term is seven times the next term. Using the trick explained, we let x equal the series, then multiply each side by 7:

$$7x = 700 + \frac{100}{7} + \frac{100}{7\cdot7} + \ldots$$

This series, after 7, is the original series. Therefore $7x = 700 + x$, or $6x = 700$, and $x = 116\frac{2}{3}$, the number of yards Achilles travels.

The bouncing ball comes to rest after traveling a distance equal to the first foot that it falls, plus the sum of 2/3 + 2/9 + 2/27 + ... The same procedure is applied (multiplying by the constant factor of 3) to obtain a limit of one foot for the series. Thus the total distance traveled by the ball, before it comes to rest after an infinite number of bounces, is 1 + 1, or two feet.

The Hungarian problem of the colored squares calls for the limit of the following series:

$$\frac{1}{9} + \frac{8}{9^2} + \frac{8^2}{9^3} + \frac{8^3}{9^4} + \cdot \cdot \cdot$$

This is also a geometric progression, with each term 9/8 of the next one. As before, we can use the algebraic trick, or—what amounts to the same thing—use the following formula for the sum of a converging series in geometric progression:

$$\frac{rx}{x - 1}$$

where r is the ratio of adjacent terms (in this case 9/8) and x is the largest term of the series (in this case 1/9). The limit is 1. Therefore as the number of coloring operations increases without bound, the colored area of the unit square approaches the area of 1. In other words, the limit is a fully covered square. Of course this could be achieved in practice only if a coloring procedure could be devised in which the time required for each step would decrease in a converging series.

The colored-squares problem was taken from *Hungarian Problem Book I*, translated by Elvira Rapaport, in the Random House New Mathematical Library (New York: Random House, 1963).

References

An Introduction to the Theory of Infinite Series. T. J. I'a. Bromwich. London: Macmillan and Co., Limited, 1942.

Summation of Infinitely Small Quantities. I. P. Natanson. Boston: D. C. Heath and Co., 1963.

Limits. Norman Miller. Waltham, Mass.: Blaisdell Publishing Co., 1964.

Limits and Continuity. William K. Smith. New York: Macmillan, 1964.

A Concept of Limits. Donald W. Hight. New York: Prentice-Hall, 1966.

Infinite Series. Earl D. Rainville. New York: Macmillan, 1967.

Modern Science and Zeno's Paradoxes. Adolf Grünbaum. Middletown, Conn.: Wesleyan University Press, 1967.

18. Polyiamonds

IN 1965 CHARLES SCRIBNER'S SONS published *Polyominoes*, a book of great interest to mathematics puzzlers. The author is Solomon W. Golomb, a mathematician then associated with the California Institute of Technology's Jet Propulsion Laboratory and professor of engineering and mathematics at the University of Southern California. It was in 1953 that Golomb, a student at Harvard University, coined the term "polyomino" for any flat figure formed by joining unit squares along their edges. Since a "domino" consists of two attached squares, Golomb proposed calling a three-square figure a "tromino," a four-square figure a "tetromino" and so on.

Among puzzle fans the 12 pentominoes— all the different ways of uniting five unit squares—proved the most popular. So intriguing were the combinatorial problems posed by these 12 little shapes that working with them became something of a national pastime. Sets of plastic pentominoes were marketed both in this country and in Britain, and Golomb found himself swamped with suggestions for new problems and requests for more information. Then, to the delight of all pentomino buffs, he assembled in one profusely illustrated volume everything of interest he had learned about the pentominoes and their square-cornered cousins.

In this chapter we consider a triangular cousin. It is mentioned briefly in Golomb's book and there are scattered references to it in a few journals, but most of what is known about this new recreation has been discovered since 1965. It is a field with many fundamental problems yet to be solved and a rich supply of patterns and theorems still to be discovered.

Golomb had pointed out as early as 1954, in "Checkerboards and Polyominoes," in *The American Mathematical Monthly*, December, 1954, that a recreation similar to polyominoes could be based on pieces formed by joining unit equilateral triangles. The Glasgow mathematician T. H. O'Beirne, writing in the *New Scientist* in 1961, proposed calling such shapes "polyamonds."

Taking his etymological cue from Golomb, O'Beirne reasoned that if a "diamond" consists of two attached triangles, a figure formed by three triangles should be called a "triamond," four triangles a "tetriamond" and so on up through "pentiamond," "hexiamond," "heptiamond" and higher *n*-iamonds. Obviously there is only one form of diamond and triamond, and the reader can quickly convince himself that there are three tetriamonds and four pentiamonds. (As with polyominoes, mirror reflections of asymmetrical forms are not usually considered different.) The hexiamonds, by a pleasing coincidence with the pentominoes, are exactly 12 in number. There are 24 heptiamonds, 66 octiamonds, and 160 order-9 figures (one with a hole). Beyond this no accurate counts have been established.

The 12 hexiamonds are shown in Figure 117. with appropriate names, most of them first proposed by O'Beirne. The reader is invited to copy these 12 shapes on a sheet of cardboard and carefully cut them out. The coloring on the shapes should be ignored. It is best to use cardboard that is the same on both sides, so that asymmetrical pieces can be turned over at will. It is good to have a supply of isometric paper on hand for ease in recording patterns.

It is obvious that any pattern formed by two or more hexiamonds must contain a number of unit triangles that is evenly divisible by 6. We can go further. By coloring the pieces as shown we see that every piece except the last two (sphinx and yacht) are "balanced" in the sense that they contain

three triangles of each color. Therefore any figure made by fitting together two or more balanced hexiamonds must itself be balanced. The yacht and sphinx are each unbalanced four to two. If one of these pieces appears in a figure, the figure must be unbalanced by an excess of two triangles. If both pieces are used, the figure must be either balanced (the yacht and sphinx being so placed that they compensate for each other) or unbalanced with an excess of four triangles. This provides a powerful check for eliminating many figures that otherwise might be thought possible.

Consider, for example, the equilateral triangle of order-6 [*Figure 118, top*]. It contains 36 unit triangles; it is the only triangle within the range of the 12 hexiamonds that has a number of unit triangles evenly divisible by 6. One could waste many hours vainly trying to construct this triangle with six hexiamonds. If it is colored as shown, however, we find that it contains an excess of six triangles of one color. Since the maximum achievable excess is four, the figure is seen at once to be impossible.

Attention turns naturally to the parallelograms. Only the 3×3 and 6×6 diamonds (rhombi) contain the proper number of triangles. The smaller diamond is easily found to be impossible, but the 6×6 has scores of

PISTOL

RHOMBOID

CLUB

HEXAGON

BUTTERFLY

SNAKE

LOBSTER

SHOE

CROWN

BAT

YACHT

SPHINX

118. Three "impossible" hexiamond patterns

known solutions. One solution, by Maurice J. Povah of Blackburn, England, is shown in Figure 120, top. It is interesting on two counts: all pieces except the hexagon touch the border, and a line divides the pattern into congruent halves. The halves can, of course, be fitted together in other ways to make bilaterally symmetrical figures.

Among the rhomboids (parallelograms with oblique angles and unequal adjacent sides) these facts are known:

1. If one side is 2, the other side must be a multiple of 3. The 2×3 is impossible. The 2×6 has one solution (ignoring independent reflections of the two halves), shown in Figure 119. It is easy to prove that only these four pieces are usable in any rhomboid with a side of 2. The rhomboidal piece leaves a space alongside it that cannot be filled, and each of the other pieces divides the figure into two areas, both of which contain an odd number of unit triangles. Since an odd number cannot be a multiple of 6, no other rhomboid with a side of 2 is possible.

2. If one side is 3, the rhomboid will contain a multiple of six triangles. The 3×3 is impossible. The $3 \times 4, 5, 6, 7, 8, 9$, and 10 are all possible, each with many solutions.

The 3×11 is possible, but it is so difficult to achieve that I leave this as an advanced exercise for the reader. In all known solu-

119. The only possible rhomboid with a side of 2

tions (one is given in the answer section) the bat is the piece left out.

The 3 × 12, which calls for all 12 hexiamonds, is the outstanding unsolved problem in the field. [*See Figure 120, bottom.*] No solution has been found, nor has an impossibility proof been devised. Can any reader cast light on this problem?

3. If one side is 4, the other must be a multiple of 3. The 4 × 3 (mentioned earlier

120. Parallelograms involving all 12 hexiamonds

as 3 × 4) is possible. So is the 4 × 6. The 4 × 9, which uses all 12 pieces, has many solutions, one of which is shown in Figure 120, middle. The shaded sections can be reflected to give three other solutions.

4. If one side is 5, the only rhomboid with a suitable number of triangles is the 5 × 6. There are many solutions.

Charles H. Lewis of Roslyn, New York, was the first to propose ring-shaped figures such as the two in Figure 118, center and bottom. It is easy to show that the triangular ring is impossible by coloring it and observing that it is unbalanced by six triangles. The hexagonal ring is balanced, but a simple impossibility proof was discovered by Meredith G. Williams of Washington, D.C. The hexagon can go in only two positions, all others being derived by rotating or reflecting the figure. In either position it is impossible to add the lobster without dividing the remaining field into two regions, neither of which has an area that is a multiple of 6.

Many patterns with threefold symmetry have been constructed. Hexagons of order-2 and order-3 exist, as is evident from the order-3 hexagon found by Adrian Struyk of Paterson, New Jersey. [see "a" in Figure 121]. Struyk also found several ways to

make the trefoil shape shown in b in the illustration. This arrangement permits the moving of one hexagon to make a straight chain of three joined hexagons. In c Struyk has bisected the trefoil into congruent halves, and in d he has produced a pattern that can be folded around a regular octahedron. Figure 122 features a variety of striking hexiamond patterns, of bilateral and threefold symmetry, discovered by Povah. Note that the figure at top right contains a solution to the problem of forming three congruent shapes using all 12 pieces.

The duplication problem—forming twice-as-high replicas of each hexiamond by using four pieces—is easily solved for each figure. As Lewis has pointed out, the two halves of the 6 × 2 rhomboid [see Figure 119] can be fitted together in various ways to duplicate all hexiamonds except the pistol, crown and lobster. The triplication problem—forming larger replicas with nine pieces—cannot be solved for the sphinx and yacht, which are unbalanced by six triangles. The other pieces are balanced, and triplications have been found for all except the butterfly. The butterfly is believed to be impossible.

Figure 123 is Povah's solution to what is called the "three twins" problem. Figure 124 shows a six-pointed star that has an eight-piece solution believed to be unique. It is not difficult, and solving it is an excellent introduction to the pleasures of hexiamondry. Here is a hint: Neither the snake, the hexagon, nor the crown can contribute to the star's perimeter.

121. Hexiamond patterns made by Adrian Struyk. Bottom pattern covers a regular octahedron.

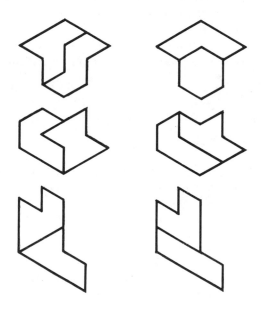

← *122. Symmetrical hexiamond patterns
by Maurice J. Povah*

123. A solution to the "three twins" problem

124. A star to be made with eight pieces

grade student in Birmingham, Alabama, was able to prove that the six-pointed star [*Figure 125*] is indeed the unique solution. All these proofs are of the exhaust-all-possibilities type and are too lengthy to give here.

The outstanding unsolved hexiamond problem — the 3-by-12 rhomboid — was solved at the Lawrence Radiation Laboratory of the University of California. A computer program written by John G. Fletcher had previously been set up for testing pentomino problems. A trivial modification by Fletcher converted this program to one capable of testing hexiamond patterns. The 3-by-12 rhomboid was found to be impossible, and the 3-by-11 rhomboid was shown to have 24 distinct solutions, all of which omit the bat. [*A solution is shown in Figure 126.*]

An earlier computer program by Mrs.

Answers

Daniel Dorritie of Endicott, New York, was the first to supply a proof that the triplication problem for the butterfly is impossible. Similar proofs were found by Esther Blackburn of Montreal; Wade E. Philpott of Lima, Ohio; and Dennis C. Rarick, a student at Indiana University. Karl Schaffer, a ninth-

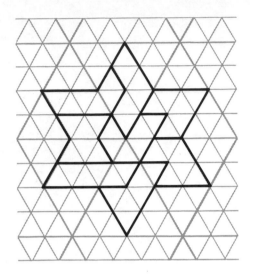

125. The only solution for the star

John Leech, in England, found 155 solutions for the 6-by-6 rhombus, 74 solutions for the 4-by-9, none for the 3-by-12. Her program, like Fletcher's, was a modification of a previous program for pentominoes. Andrew L. Clarke, Wellesey, England, supplied proofs (without computer aid) for the 3-by-12 rhombus, the butterfly, and the six-pointed star.

Sets of plastic hexiamonds were on sale in the late 1960's, under various trade names, in England, Japan, and West Germany.

References

"Maestro Puzzles." J. E. Reeve and J. A. Tyrrell. *The Mathematical Gazette*, Vol. 45, No. 352; May, 1961. Pages 97–99.

"Pentominoes and Hexiamonds." T. H. O'Beirne. *New Scientist*, Vol. 12, No. 259; November, 1961. Pages 316–317.

"Some Hexiamond Solutions: and an Introduction to a Set of 25 Remarkable Points." T. H. O'Beirne. *New Scientist*, Vol. 12, No. 260; November, 1961. Pages 379–380.

"Thirty-Six Triangles Make Six Hexiamonds Make One Triangle." T. H. O'Beirne. *New Scientist*, Vol. 12, No. 265; December, 1961. Pages 706–707.

Die 12 Verhext. Herbert Zimpfer, Baden: privately printed, 1967.

"Polyiamonds." Ir. P. J. Torbijn. *Journal of Recreational Mathematics*, Vol. 2, No. 4; October, 1969. Pages 216–227.

126. Forming the 3-by-11 rhomboid

19. Tetrahedrons

ANY FOUR POINTS (A, B, C, D) in space that are not all on the same plane mark the corners of four triangles [*see Figure 127*]. These triangles in turn are the faces of a tetrahedron, the simplest of all polyhedrons (solids bounded by polygons). If each face of a tetrahedron is an equilateral triangle, it is a regular tetrahedron, the simplest of the five platonic solids. Indeed, it is so simple that it was known in ancient Egypt and was probably studied by mathematicians as early as the cube.

The Greek Pythagoreans believed that fire was composed of tetrahedral particles too small to be seen. Because the tetrahedron has fewer faces and sharper corners than any other regular convex solid, they argued, tetrahedral particles would form the least stable and most "penetrating" of the four elements: earth, air, fire and water. We know better today, yet there is a sense in which this Pythagorean guess, like so many guesses of that school, was a shrewd one, for the tetrahedral structure does turn up in many aspects of the microworld. The

so-called carbon atom, without which organic molecules and life as we know it would not be possible, is actually an atom of carbon joined by chemical bonds to four other atoms vibrating at the vertices of a tetrahedron. For example, a molecule of carbon tetrachloride, the familiar cleaning fluid, consists of one carbon atom bonded in this way to four atoms of chlorine. Many crystal lattices, including that of diamond, have a tetrahedral structure. An important copper ore that has a tetrahedral lattice is called tetrahedrite because it is found so often in large, well-developed tetrahedral crystals.

Squares of the same size fit together like a checkerboard to fill the plane, and in a similar way cubes join to fill space. Because equilateral triangles also tile a plane, one might suppose that regular and congruent tetrahedrons would also pack snugly to fill space. This seems so intuitively evident that even Aristotle, in his work *On the Heavens*, declared it to be the case. The fact is that among the platonic solids the

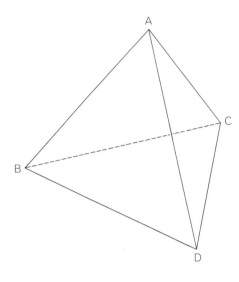

127. *A regular tetrahedron*

cube alone has this property. If the tetrahedron also had it, it would long ago have rivaled the cube in popularity for packaging.

Interestingly, regular tetrahedrons and octahedrons (regular solids bounded by eight triangles) *will* pack to fill space if they are arranged alternately as shown in Figure 128. They are the only two regular solids that fit together to fill space. Note that every triangle in the lattice is the face of both a tetrahedron and an octahedron, and that every vertex is surrounded by eight tetrahedrons and six octahedrons. This beautifully regular structure has been exploited in recent years by the inventor-architect R. Buckminster Fuller. The canti-levered truss he calls the "octet" consists of aluminum tubing joined in a network that traces the edges of an octahedral-tetrahedral honeycomb. (A stimulating classroom project is to model such a honeycomb by joining the ends of a large number of rods

or soda straws that are all the same length.) Fuller's more famous "geodesic" domes are essentially tetrahedral lattices intended, like his octet, to achieve maximum rigidity at minimum weight and cost.

Fuller is not the first well-known American inventor to be fascinated by the tetrahedron's great strength-to-weight ratio. After Alexander Graham Bell achieved fame as the inventor of the telephone he developed an almost obsessive interest in tetrahedrons. Efforts to build airplanes in the 1890's had failed because engines lacked the power to keep the craft airborne, and Bell decided that the answer lay in constructing enormous silk-covered, man-carrying kites honeycombed with a tetrahedral lattice of aluminum tubing. At his summer home in Baddeck, Nova Scotia, he built and flew a fantastic variety of such kites. To observe his kites in flight he had an 80-foot-high platform constructed at the top vertex of a tetrahedral skeleton formed by three trusses, each of which was a tetrahedral network. On the ground he built a wooden observation hut also shaped like a tetrahedron. When the Alexander Graham Bell Museum was built at Baddeck in 1955, a tetrahedral pattern was used throughout the building as a basic architectural motif.

Bell would surely have been delighted by recent adaptations of the tetrahedral shape to packaging. If you pinch together the bottom of a paper tube and tape it to form a straight edge, then do the same thing at the top of the tube but at right angles, a tetrahedron results. If the tube's circum-

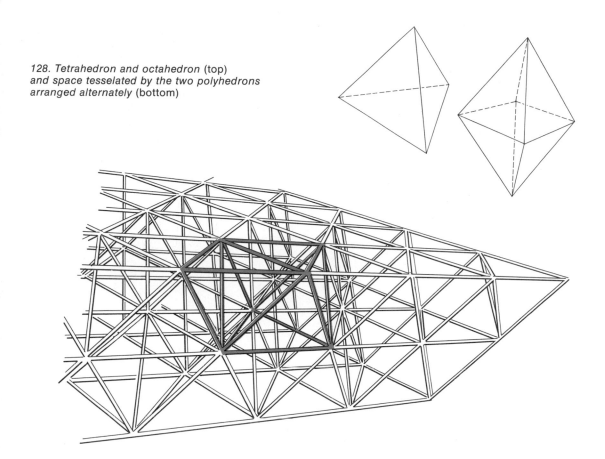

128. *Tetrahedron and octahedron* (top)
and space tesselated by the two polyhedrons
arranged alternately (bottom)

ference is 4 units and its height is the square root of 3, the tetrahedron will be regular [*Figure 129*]. This efficient method of construction underlies Tetra Pak, the trade name for a paper container developed in Sweden in the mid-1950's. It first swept through Europe and is now being used increasingly in the U.S., chiefly as a milk container and coffee creamer.

A quite different application of the tetrahedral shape is shown in Figure 130. Dur-ing World War II the four-pronged device called a "caltrop" (it might be interpreted as a model of the carbon atom!) was used for puncturing the tires of enemy vehicles. Hundreds of them can be tossed along a road and every one will land with one spike pointing straight up; moreover, the shape permits maximum penetration of a tire. The idea is an old one. *The Oxford English Dictionary* defines a caltrop as "an iron ball armed with four sharp prongs or spikes,

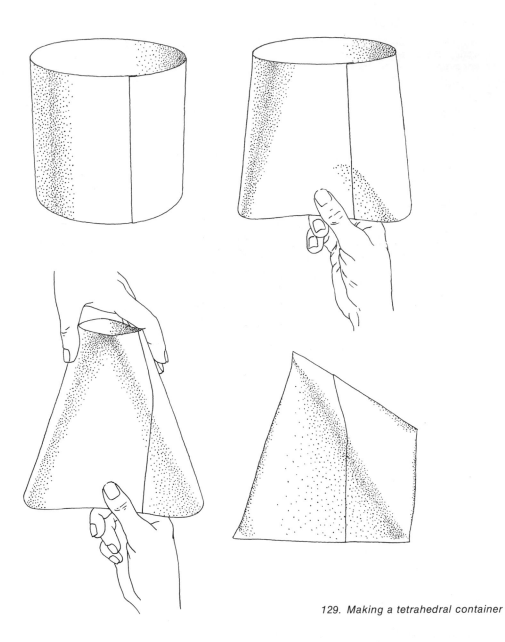

129. Making a tetrahedral container

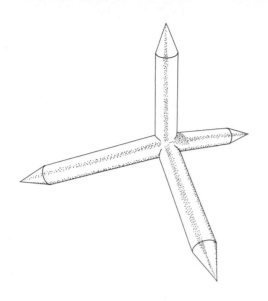

130. Tetrahedral tire-puncturing device

efficient breakwaters. The *New York Times*, February 21, 1965, page S19, described their widespread use at the Port of Ashdod in Israel.

The four-dimensional analogue of the tetrahedron is called a pentatope. If a point at the center of an equilateral triangle is joined to each vertex, the result is a projection on the plane of a tetrahedron's skeleton. In similar fashion we can join a point at the center of a tetrahedron to the four vertices and obtain a projection in three-space of the skeleton of a pentatope [*Figure 131*]. It is easy to see that the pentatope has five vertices, ten edges, ten triangular faces

131. Projection in three-space of a pentatope

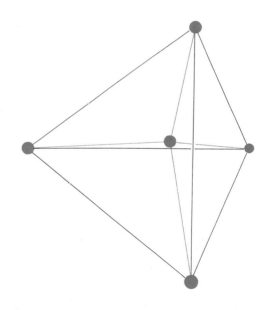

placed like the angles of a tetrahedron, so that when thrown on the ground it has always one spike projecting upwards: Used to obstruct the advance of cavalry, etc." One of the dictionary's several quotations from sixteenth century documents reads: "The Irishmen had strawed all alongst the shore a great number of caltrops of iron, with sharp pricks standing up, to wound the Danes in the feet." And Oliver Wendell Holmes, in 1858, wrote: "One of those small *calthrops* our grandfathers used to sow round in the grass when there were Indians about . . ."

A more recent use for the caltrop structure is provided by the "tetrapod," a monstrous four-limbed object made of reinforced concrete and weighing many tons. It resembles a fat caltrop with flat instead of pointed ends. When thousands are piled together on a beach, they interlock to provide highly and five tetrahedral cells. (In this projection we see four small cells and one large one. On the pentatope itself, if it is regular, all five cells are congruent.) Any five points in four-dimensional space that are not on the same three-space hyperplane mark the corners of a pentatope, and each set of four points establishes the corners of a tetrahedral cell. If the five points are so placed in four-space that each pair is the same distance apart, the figure is a regular pentatope, one of the six regular convex solids of the fourth dimension.

Just as a tetrahedron's four faces can be unfolded to make a plane figure consisting of a central triangle with a triangle attached to each edge, so the five tetrahedral cells of a pentatope that form its hypersurface can be "unfolded" into three-space to make a stellated tetrahedron: a central tetrahedron with a tetrahedron on each face [*see Figure 132*]. If we only knew how to fold such a solid through the fourth dimension, we could fold it into a pentatopal container for hypercream.

A strange, little-known property of the regular tetrahedron—a property it does not share with any other platonic solid—is involved in a perplexing magic trick that can be presented as a demonstration of one's ability to sense color vibrations with the fingers. First construct a small model of a regular tetrahedron, its faces congruent with the triangles in Figure 133. (A quick way to make such a model has been proposed by Charles W. Trigg. Cut the pattern shown in Figure 134 from stiff paper or light cardboard. Crease all lines the same

way, fold the white triangles into a tetrahedron, then tuck the shaded triangles into open edges to form a stable, no-paste-required model.) Place the model on the black triangle at the top of the pattern (or on a board made by copying the pattern with different colors for each of the numbered shades). While your back is turned, someone "rolls" the model at random over the pattern by tipping it over an edge from triangle to triangle. He stops whenever he pleases, notes the color on which it rests and lets it remain there while he counts slowly to 10. Then he *slides* the tetrahedron back to the black triangle. You turn around, pick up the tetrahedron, feel its underside and name the color on which it last rested.

The secret combines geometry with a card hustler's dodge. A common method of marking a deck of cards while a game is in progress is to obtain a smear of what is called "daub" on the tip of a finger, then press it to the margin of a card at a spot that codes the card's value. The daub leaves only a dim smudge, indistinguishable from the kind of dirt marks that normally dull the margins of cards that have been much used. Make some daub by rubbing a pencil point heavily over the same spot on a piece of paper. Slide a fingertip over the graphite, then press the tip lightly against the corner of one face of the tetrahedron. The idea is to leave such a faint smudge that no one but you will ever notice it.

Place the secretly marked tetrahedron on the black triangle with the mark at the top corner and facing the pattern. At the end of the trick the location of the smudge

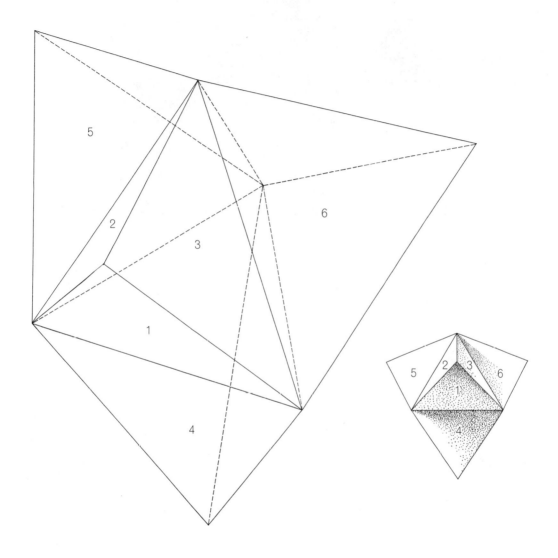

132. *Pentatope unfolded into three-space*

133. Board pattern for the magic trick

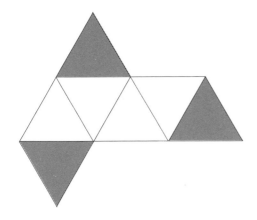

134. Pattern for folding a tetrahedral counter

will code the color on which the model last rested. As you pretend to feel the base of the model, look directly down at it. The smudge will be at one of four positions, each of which indicates a different color [see Figure 135]. I leave it to the reader to discover why the trick cannot fail.

The following puzzles involving tetrahedrons are not difficult, but some have surprising solutions.

1. A regular tetrahedron is cut simultaneously by six different planes. Each slices the solid exactly in half by passing through one edge and bisecting the opposite edge. How many pieces result?

2. Can any triangle cut from paper be folded along three lines to form a (not necessarily regular) tetrahedron? If not, give the conditions that must be met.

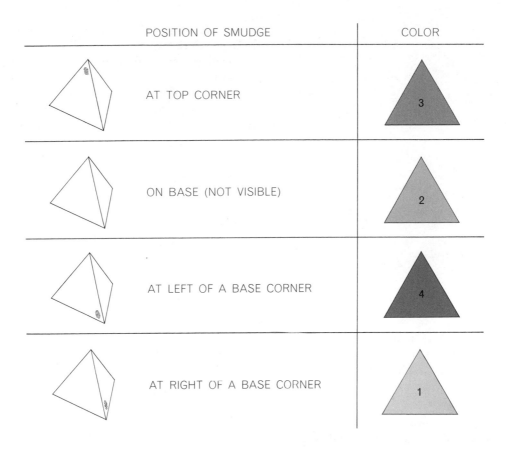

POSITION OF SMUDGE	COLOR
AT TOP CORNER	3
ON BASE (NOT VISIBLE)	2
AT LEFT OF A BASE CORNER	4
AT RIGHT OF A BASE CORNER	1

135. The key to the magic trick

3. Inside a room shaped like a regular tetrahedron a bug crawls from point *A* to point *B* [*see Figure 136*]. The room is 20 feet on a side and each point is seven feet from a vertex, on an altitude of a triangular wall. What is the length of the bug's shortest path?

4. What is the largest number of spots that can be painted on a sphere so that the distance between every pair is the same?

5. If a regular tetrahedron one inch on a side is cut from each corner of a tetrahedron with a side of two inches, what kind of solid is left?

6. Is it possible to label each face of a tetrahedron with a different number so

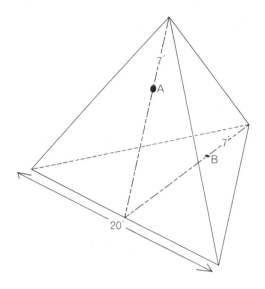

136. The bug problem

that the sum of the three faces meeting at each vertex is the same? Is it possible to label each *edge* so that the sum of the three edges meeting at each vertex is the same? In both cases numbers may be rational or irrational.

7. What is the length of the side of the largest regular tetrahedron that can be packed into a cubical space one foot on a side?

8. How many different tetrahedrons can be made by joining four equilateral cardboard triangles each of which has a different color? Two tetrahedrons are considered alike only if one can be turned and placed beside the other so that the color patterns of the two figures match. If the patterns can be made to match only by mirror reflection, they are considered different.

9. If each side of a regular tetrahedron is painted either red or blue, it is easy to see that only five different models can be made: one all red, one all blue, one with one red side, one with one blue side, and one with two red and two blue sides. If each side is painted either red, white, or blue, how many different models can be made? As before, rotations are not regarded as different.

Answers

1. A regular tetrahedron cut by six planes, each passing through an edge and bisecting the opposite edge, will be sliced into 24 pieces. This is easily seen when one realizes that each face is dissected into six triangles, as in *a* in Figure 137, each of which is the base of a tetrahedron with its apex at the model's center. (This problem was contributed by Harry Langman to *Scripta Mathematica* for March–June, 1951.)

2. Any paper triangle, if all its angles are acute, can be folded into a tetrahedron.

3. The bug's shortest path from *A* to *B* is 20 feet, as shown on the unfolded tetrahedron at *b* in Figure 137. This is shorter by .64+ feet than the shortest path that does not touch a third face.

4. Four is the largest number of spots that can be placed on a sphere so that every pair is the same distance apart. The spots mark the corners of an inscribed regular tetrahedron.

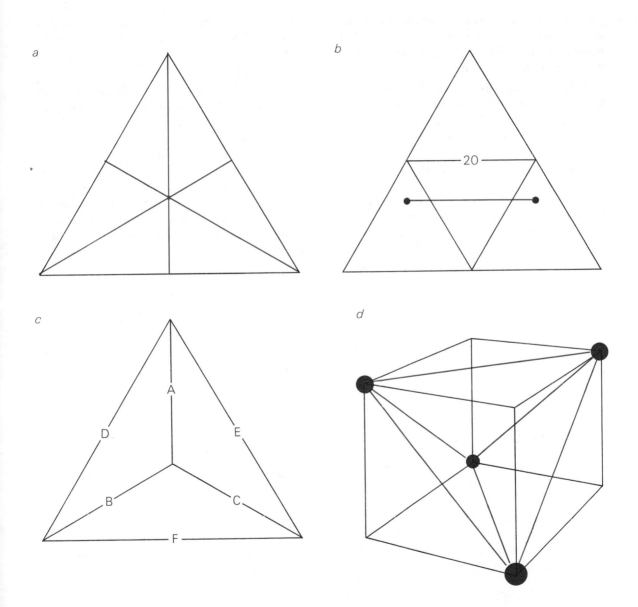

a

b

20

c

A

D   E

B   C

F

d

137. Answers to the tetrahedron problems

5. If one-inch regular tetrahedrons are sliced from the four corners of a two-inch regular tetrahedron, the remaining solid is a regular octahedron.

6. It is not possible to label the sides of a tetrahedron with four different numbers so that the sum of the three faces at each vertex is the same. Consider any two sides A and B. They meet side C at one vertex and side D at another. For the sums at both vertices to be constant the numbers on sides C and D would have to be the same, but this violates the condition that the four numbers must be different.

A proof (from Leo Moser) that the edges of a tetrahedron cannot be labeled with six different numbers to yield constant corner sums is a bit more involved. First label the edges as shown at c in Figure 137. Assume that the problem can be solved. Then $a + b + c = a + e + d$, therefore $b + c = e + d$. Similarly, $f + b + d = f + e + c$, therefore $b + d = e + c$. Add the two equations:

$$\begin{array}{r} b + c = e + d \\ b + d = e + c \\ \hline 2b + c + d = 2e + c + d. \end{array}$$

The sum reduces to $b = e$, which of course violates the assumption that no two numbers are the same.

7. The largest regular tetrahedron that can be placed inside a unit cube has a side the length of which is the square root of 2 [“d” in Figure 137].

8. Four equilateral cardboard triangles of four different colors will combine to make two different tetrahedrons, one a mirror image of the other.

9. If each side of a regular tetrahedron is painted red, white or blue, it is possible to paint 15 different models: three will be all one color, three will have red-blue faces, three will have red-white faces, three will have blue-white faces, and three will have red-white-blue faces with two faces of the same color. The formula for the number of different tetrahedrons (counting mirror reflections but not rotations as being different) that can be made with n colors is

$$\frac{n^4 + 11n^2}{12}.$$

References

“The Tetrahedral Principle in Kite Structure.” Alexander Graham Bell. *National Geographic*, Vol. 14, No. 6; June, 1903. Pages 219–251.

“Geometry of Paper Folding, II. Tetrahedral Models.” C. W. Trigg. *School Science and Mathematics*, Vol. 54, December, 1954. Pages 683–689.

“Alexander Graham Bell Museum: Tribute to Genius.” Jean Lesage. *National Geographic*, Vol. 60, No. 2; August, 1956. Pages 227–256.

Regular Polytopes. H. S. M. Coxeter. New York: The Macmillan Company, 1963.

20. Coleridge's Apples and Eight Other Problems

1. Coleridge's Apples

Who would have thought that the poet Samuel Taylor Coleridge would have been interested in recreational mathematics? Yet the first entry in the first volume of his private notebooks (published in 1957 by Pantheon Books) reads: "Think any number you like—double—add 12 to it—halve it—take away the original number—and there remains six." Several years later, in a newspaper article, Coleridge spoke of the value of this simple trick in teaching principles of arithmetic to the "very young."

The notebook's second entry is: "Go into an Orchard—in which there are three gates—thro' all of which you must pass—Take a certain number of apples—to the first man [presumably a man stands by each gate] I give half of that number & half an apple—to the 2nd [man I give] half of what remain & half an apple—to the third [man] half of what remain & half an apple—and yet I never cut one Apple."

How long will it take the reader to deter-mine the smallest number of apples Coleridge could start with and fulfill all the stated conditions?

2. Reversed Trousers

Each end of a 10-foot length of rope is tied securely to a man's ankles. Without cutting or untying the rope, is it possible to remove his trousers, turn them inside out on the rope and put them back on correctly? Party guests should try to answer this confusing topological question before initiating any empirical tests.

3. Coin Game

The two-person game shown in Figure 138 has been designed to illustrate a principle that is often of decisive importance in the end games of checkers, chess, and other mathematical board games. Place a penny on the spot numbered 2, a dime on spot 15.

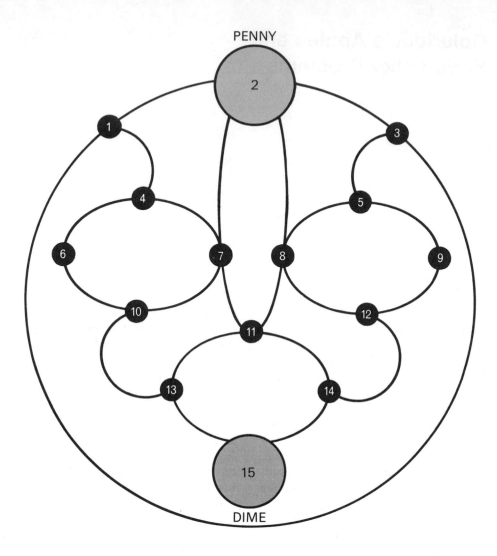

PENNY

DIME

138. Can the penny always trap the dime?

Players alternate turns, one moving the penny, the other the dime. Moves are made along a solid black line to an adjacent spot. The penny player always moves first. His object is to capture the dime by moving onto the spot occupied by the dime. To win he must do so before he makes his seventh move. If after six of his moves he has failed to catch the dime, he loses.

There is a simple strategy by which one player can always win. Can the reader discover it?

4. Truthers, Liars, and Randomizers

Logic problems involving truth-tellers and liars are legion, but the following unusual variation—first called to my attention by Howard De Long of West Hartford, Connecticut—had not to my knowledge been printed before it appeared in *Scientific American*.

Three men stand before you. One always answers questions truthfully, one always responds with lies and one randomizes his answers, sometimes lying and sometimes not. You do not know which man does which, but the men themselves do. How can you identify all three men by asking three questions? Each question may be directed toward any man you choose, and each must be a question that is answered by "Yes" or "No."

5. Gear Paradox

The mechanical device shown in Figure 139 was constructed by James Ferguson, an eighteenth-century Scottish astronomer well known in his time as a popular lecturer, author and inventor, and for the remarkable fact that although he was a member of the Royal Society his formal schooling had consisted of no more than three months in grammar school. (One of his biographies is called *The Story of the*

139. James Ferguson's gear paradox

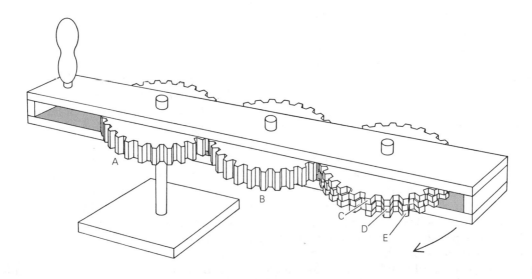

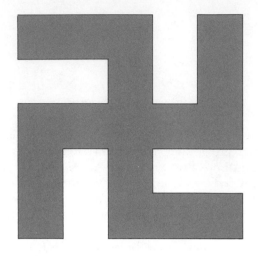

140. Model for the swastika

Peasant-Boy Philosopher.) His device is given here as a puzzle that, once solved, will be seen to be a most curious paradox.

Wheel *A* and its axis are firmly fixed so that wheel *A* cannot turn. When the device is rotated clockwise around wheel *A* by means of the handle, wheel *B* will of course rotate in the same direction. The teeth of *B* engage the teeth of three thinner wheels *C*, *D*, and *E*, each of which turns independently. *A*, *B*, and *E* each have 30 teeth. *C* has 29 teeth, *D* has 31. All wheels are of the same diameter.

As seen by someone looking down on the device as it is turned clockwise, each of the thin wheels *C*, *D*, and *E* must turn on its axis (with respect to the observer) either clockwise, counterclockwise, or not at all. Without constructing a model, describe the motion of each wheel. If the reader wishes to build a model eventually, it is not necessary that the wheels have the exact number of teeth given. It is only necessary that *A* and *E* have the same number of teeth, *C* at least one less, and *D* at least one more.

6. Form a Swastika

During World War II a gag problem that made the rounds was: How can you make a Nazi cross with five matches? One answer was "Push four of them up his rear end and light them with the fifth." Here is a somewhat similar problem, although one that does not hinge on wordplay.

The reader is asked to take four cigarettes

and eight sugar cubes, place them on a dark-surfaced table top and form the best possible replica of the swastika (a mirror image of the Nazi symbol) shown in Figure 140. All 12 objects must be used and none must be damaged in any way.

7. Blades of Grass Game

According to a recent book by two Soviet mathematicians, the following method of fortune-telling was once popular in certain rural areas of the U.S.S.R. A girl would hold in her fist six long blades of grass, the ends protruding above and below. Another girl would tie the six upper ends in pairs, choosing the pairs at random, and then tie the six lower ends in a like manner. If this produced one large ring, it indicated that the girl who did the tying would be married within a year.

A pencil-and-paper betting game (a pleasant way to decide who pays for drinks) can be based on this procedure. Draw six vertical lines on a sheet of paper. The first player joins pairs of upper ends in any manner, then folds back the top of the paper to conceal the connecting lines from his opponent. The second player now joins pairs of bottom ends as shown at the left in Figure 141. The sheet is unfolded to see if the second player has won by forming one large closed loop. (The illustration at the

right of Figure 141 shows such a win.) If even money is bet, whom does the game favor and what is his probability of winning?

8. Casey at the Bat

During a baseball game in Mudville, Casey was Mudville's lead-off batter. There were no substitutions or changes in the batting order of the nine Mudville men throughout

141. *A pencil-and-paper betting game*

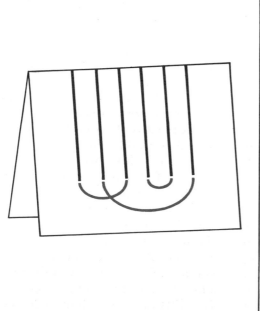
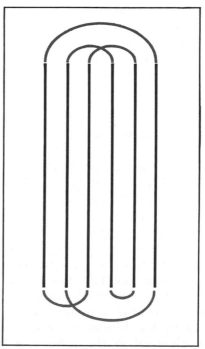

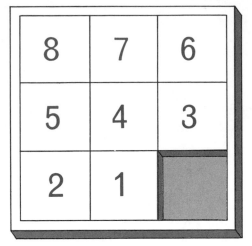
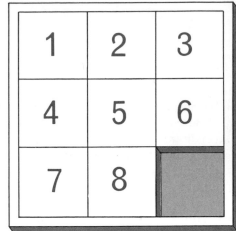

142. Change the pattern at the left to the one at the right

the nine-inning game. It turned out that Casey came to bat in every inning. What is the least number of runs Mudville could have scored? Charles Vanden Eynden of the University of Arizona originated this amusing problem.

9. The Eight-Block Puzzle

Sam Loyd's well-known 14–15 Puzzle was mentioned in this book's chapter on sliding-block puzzles. For all puzzles of this type, in which unit squares are shifted about inside a rectangle by virtue of a "hole" that is also a unit square, there is a quick parity check for determining if one pattern can be obtained from another. For example, on the simplest nontrivial square field shown in Figure 142 can the pattern at the left (with the blocks in descending order) be changed to the pattern at the

right? To answer this we switch pairs of numbers (by removing and replacing blocks) until the desired pattern is achieved, counting the switches as we go along. This can be done in helter-skelter fashion, with no attempt at efficiency. If the number of switches is even (as it always will be in this case), the change by sliding is possible. Otherwise it is not.

But what is the *smallest* number of sliding moves sufficient to make this change? Surprisingly little work has been done on methods for minimizing such solutions. The problem given here—reversing the order of the digits—can be shown to require at least 26 moves. If each square takes the shortest path to its destination, 16 moves are used. But 4 and 5 are adjacent and cannot be exchanged in fewer than four moves, and the same reasoning applies to 3 and 6. This lifts the lower limit to 20. Two moves are lost by the opening move of 1 or 3 and two more by the last move of 8 or 6, since in each case a square must occupy a cell outside its shortest path. This raises the lower limit to 24. Finally, if one constructs a tree graph for opening lines of play, it is apparent that two more moves must be lost by the ninth move. The puzzle therefore cannot be solved in fewer than 26 moves. Because the hole returns to its original position, it can be shown that every solution will have an even number of moves.

The best solution on record (it is the solution to problem 253 in Henry Ernest Dudeney's posthumous collection, *Puzzles and Curious Problems*) requires 36 moves. Recorded as a chain of digits to show the order in which the pieces are moved, it is as follows: 12543 12376 12376 12375 48123 65765 84785 6. I have good reason to believe, however, that it can be done in fewer moves.

To work on the problem one can move small cardboard squares, numbered 1 through 8, on a square field sketched on paper, or work with nine playing cards on a rectangular field. I shall be grateful if readers who do better than Dudeney will send their solutions to me.

Answers

1.

Seven is the smallest number of apples (we rule out "negative apples") that satisfies the conditions of Coleridge's problem.

2.

To reverse a man's trousers while his ankles are joined by rope, first slide the trousers off onto the rope, then push one leg through the other. The outside leg is reversed twice in this process, leaving the trousers on the rope right side out but with the legs exchanged and pointing toward the man's feet. Reach into the trousers from the waist and turn both legs inside out. The trousers are now reversed on the rope and in position to be slipped back on the man, zipper in front as originally but with the legs interchanged.

3.

In analyzing the topological properties of a network with an unusual pattern it is sometimes helpful to transform the network to a topologically equivalent one that exhibits the network's regularities better. The pattern of the penny-dime game [*at top of Figure 143*] is readily seen to be equivalent to the board at the bottom in the illustration. If the penny moves directly toward the dime, it cannot trap it because the dime has what in chess and checkers is called the "opposition." The meaning of this term is brought out by coloring every other spot. As long as both pieces avoid the triangle at the upper right the dime's move will always carry it to a spot of the same color as the spot occupied by the penny; therefore the penny, on its next move, can never catch the dime. To gain the opposition the penny must move once along the

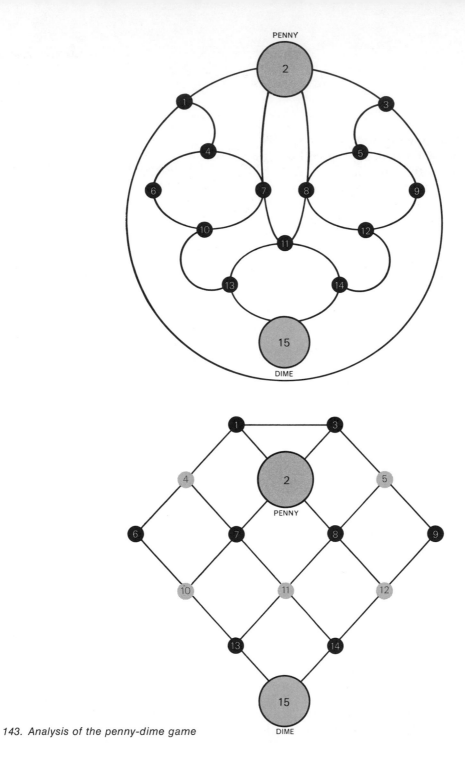

143. Analysis of the penny-dime game

long outside arc that joins the two colored spots numbered 1 and 3. Because this alters the relative parity of the two pieces it is then a simple matter for the penny to corner the dime.

Translating back to the original board, this means that the penny's best strategy is to move either first to 1, then all around the outside circle to 3, or first to 3 and then around to 1. In either case the penny will then have no difficulty trapping the dime, on spot 6, 9 or 15, before the seventh move.

4.

Label the three men, A, B, C, and let T stand for truth-teller, L for liar, and R for randomizer. There are six possible permutations of T, L, and R:

	A	B	C
(1)	T	L	R
(2)	T	R	L
(3)	L	R	T
(4)	L	T	R
(5)	R	T	L
(6)	R	L	T

Ask A "Is B more likely to tell the truth than C?" If he answers, "Yes," lines 1 and 4 are eliminated and you know that C is not the randomizer. If he answers "No," lines 2 and 3 are eliminated and you know that B is not the randomizer. In either case, turn to the man who is not the randomizer and ask any question for which you both know the answer. For example: "Are you the randomizer?" His answer will establish

whether he is the truth-teller or the liar. Knowing this, you can ask him if a certain one of his companions is the randomizer. His answer will establish the identities of the other two men.

Many readers sent different solutions. The most unusual, by Kenneth O'Toole, was passed along to me by Mary S. Bernstein. A man is asked, "If I asked each of you if I had on a hat, and your two companions gave the same answer, would your answer agree with theirs?" The truther says no, the liar yes, and the randomizer cannot reply because he knows his companions cannot agree. Here we encounter ambiguity because, in a sense, any answer by the randomizer would be a "lie." Assuming, however, that the randomizer remains silent, the question need be asked of only two men to identify all three.

5.

When James Ferguson's curious mechanical device is turned clockwise, wheel C rotates clockwise in relation to the observer, D rotates counterclockwise, and E does not rotate at all!

6.

Four cigarettes and eight sugar cubes can be placed on a dark surface to form an excellent replica of a swastika, as shown in Figure 144.

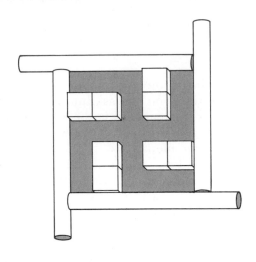

144. *Solution of the swastika puzzle*

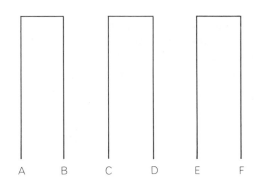

A   B   C   D   E   F

145. *The blades-of-grass problem*

7.

What is the probability of forming one ring by a random joining of pairs of upper ends of six blades of grass, followed by a random joining of pairs of lower ends? Regardless of how the upper ends are joined, we can always arrange the blades as shown in Figure 145. We now have only to determine the probability that a random pairing of lower ends will make a ring.

If end *A* is joined to *B*, the final outcome cannot be one large ring. If, however, it is joined to *C*, *D*, *E*, or *F*, the ring remains possible. There is therefore a probability of 4/5 that the first join will not be disastrous. Assume that *A* is joined to *C*. *B* may now join *D*, *E*, or *F*. Only *D* is fatal. The probability is 2/3 that it will join *E* or *F*, and in either case the remaining pair of ends must complete the large ring. The

same would hold if *A* had been joined to *D*, *E*, or *F* instead of to *C*. Therefore the probability of completing the ring is 4/5 × 2/3 = 8/15 = .53+. That the probability is better than half is somewhat unexpected. This means that in the pencil-and-paper version explained earlier the second player has a slight advantage. Since most people would expect the contrary, it makes a sneaky game to propose for deciding who picks up the tab. Of course you magnanimously allow your companion to play first.

The problem generalizes easily. For two blades of grass the probability is 1 (certain), for four blades it is 2/3, for six it is 2/3 × 4/5, and for eight blades, 2/3 × 4/5 × 6/7. For each additional pair of blades simply add another fraction, easily determined because the numerators of this series are the even numbers in sequence and the denominators are the odd numbers in sequence! For a derivation of this simple formula, and the use of Stirling's formula to approximate the probability when very large numbers of fractions must be multiplied, see *Challenging Mathematical Problems with Elementary Solutions, Vol. I,* by the Russian twin brothers A. M. and I. M. Yaglom. (It is problem No. 78 in the English translation by James McCawley, Jr.; San Francisco: Holden-Day, 1964.)

Waldean Schulz included the following additional twist to this problem in a paper titled "Brain Teasers and Information Theory" that he wrote for a philosophy class at the University of Colorado taught by David Hawkins. Suppose you are the second player. How can you join the lower ends in such a way that you can ask a single yes-no question which, if answered by the first player, will tell you if you won or lost?

The answer is to join the two outside lines, the two next-to-outside lines, and the two middle lines. The question is: Did you connect the upper ends in a bilaterally symmetric way? A yes answer means you lost, a no answer means there is a single loop and you win. It is surprising that one "bit" of information is sufficient to distinguish between winning and losing patterns.

8.

The Mudville team could have scored as few as no runs at all even though Casey, the lead-off man, came to bat every inning. In the first inning Casey and the next two batters walk and the next three strike out. In the second inning the first three men walk again, which brings Casey back to bat. But each runner is caught off base by the pitcher, so Casey is back at the plate at the start of the third inning. This pattern is now repeated until the game ends with no joy in Mudville, even though the mighty Casey never once strikes out.

There are, of course, many other ways the game could be played. Robert Kaplan, Cambridge, Massachusetts, wrote the following letter:

Dear Mr. Gardner:

That was indeed an amusing problem concerning Casey and the Mudville nine — amusing, that is, to all save lovers of Mudville. For on the unfortunate day described in your problem, Mudville scored not a run. This is what happened:

In the first inning, Casey and two of his confreres reached base, but batters four through six struck, flied, or otherwise made out. No runs.

In the second inning, batters seven and eight struck out, let us say, but the Mudville pitcher, to the surprise of all, reached base on a bobbled infield roller. Casey came up to bat, frowning mightily. With the count two and two, the perfidious rival pitcher, ignoring the best interests of poetry, baseball mythology and Mudville, whirled toward first and picked off his opposite number, who, dreaming of Cooperstown and the Hall of Fame, had strolled too far from the bag. The crowd sighed, Casey glowered, and the inning was over: no runs.

Now as you know, if an inning ends with a pick-off play at any base, the batter who was in the box at the time becomes the first batter next inning. So it was with Casey; once again Mudville loaded the bases; but once again three outs were made with no runs scoring, so that the inning ended with batter six making the last out.

Life may be linear but fate is cyclic: innings four, six, and eight followed precisely the same pattern as inning two (though you may be sure that after his second miscue in the fourth, the Mudville pitcher was lifted and his relief was responsible not only for the flood of runs the opponents scored, but for similar cloud-gazing on the base-paths). And of course, Casey led off the fifth, seventh, and ninth innings as he had the third—and again, Mudville would load the bases, but could not deliver (if I remember correctly, the gentlemen responsible for this orgy of weak hitting were Cooney, Burrows, Blake and Flynn). Grand total for Mudville: a gooseegg.

Sincerely though sorrowfully yours,
ROBERT KAPLAN

P.S. I see in rereading the problem that there were no substitutions or changes in the Mudville batting order during the game. How, then, you might justly ask, did the crowd or the manager tolerate such flagrant disdain of first base on the part of their pitcher? The answer is that he was married to the owner's daughter, and no one could say him nay.

9.

The sliding-block puzzle can be solved in 30 moves. I had hoped I could list the names of all readers who found a 30-move solution, but the letters kept coming until there were far too many names for the available space. All together readers found ten different 30-movers. They are shown paired in Figure 146, because, as many readers pointed out, each solution has its inverse, obtained by substituting for each

1a.	34785	21743	74863	86521	47865	21478
1b.	12587	43125	87431	63152	65287	41256
2a.	34785	21785	21785	64385	64364	21458
2b.	14587	53653	41653	41287	41287	41256
3a.	34521	54354	78214	78638	62147	58658
3b.	14314	25873	16312	58712	54654	87456
4a.	34521	57643	57682	17684	35684	21456
4b.	34587	51346	51328	71324	65324	87456
5a.	12587	48528	31825	74316	31257	41258
5b.	14785	24786	38652	47186	17415	21478

146. Minimum-move sliding-block solutions

digit its difference from 9 and taking the digits in reverse order. Note that of the four possible two-move openings, only 3,6 does not lead to a minimum-move solution. Solutions *2a* and *3b* proved to be the easiest to find. The most elusive solution, *5a*, was discovered by only ten readers. Only two readers, H. L. Fry and George E. Raynor, found all ten without the help of a computer.

Donald Michie of the University of Edinburgh has been using this eight-block puzzle in his work on game-learning machines. His colleague Peter Schofield, of the university's computer unit, had written a program for determining minimum solutions for all the 20,160 patterns that begin and end with the hole in the center. (Of these patterns, 60 require 30 moves, the maximum for center-hole problems.) With the aid of this program Schofield was able to find all ten solutions, but this did not rule out the possibility of others, or even of a shorter solution. The matter was first laid to rest by William F. Dempster, a computer programmer at the Lawrence Radiation Laboratory of the University of California at Berkeley, with a program for an IBM 7094. It first ran off all solutions of 30 moves or fewer, printing out the ten solutions in 2½ minutes. A second run, for all solutions of 34 moves or fewer, took 15 minutes. It produced 112 solutions of 32 moves and 512 solutions of 34 moves. There are therefore 634 solutions superior to the 36-mover given by Henry Ernest Dudeney, who first posed the problem. The ten 30-movers were later confirmed by about a dozen other computer programs. It is not yet known if there are starting and ending patterns, with the hole in the corner or side cell, that require more than 30 moves.

21. The Lattice of Integers

THE SIMPLEST of all lattices in a plane—taking the word "lattice" in its crystallographic sense—is an array of points in square formation. This is often called the "lattice of integers," because if we think of the plane as a Cartesian coordinate system, the lattice is merely the set of all points on the plane whose x and y coordinates are integers. Figure 147 shows a finite portion of this set: the 441 points whose coordinates range from 0 to 20.

Think of the 0,0 point as the southwest corner of a square orchard, fenced on its south and west sides, but infinite in its extension to the north and east. At each lattice point is a tree. If you stand at 0,0 and peer into the orchard, some trees will be visible and others will be hidden behind closer trees. Here, of course, our analogy breaks down, because the trees must be taken as points and we consider any tree "visible" to one eye at 0,0 if a straight line from that point to the tree does not pass through another point. The colored dots mark all lattice points visible from 0,0; the unmarked grid intersections represent points that are not visible.

If we identify each point with a fraction formed by placing the point's y coordinate over its x coordinate, many interesting properties of the lattice (properties first called to my attention by Robert B. Ely of Philadelphia) begin to emerge. For example, each visible point is a fraction whose numerator and denominator are coprime; that is, they have no common factor other than 1 and therefore cannot be reduced to a simpler form. Each invisible point is a fraction that *can* be simplified—and each simplification corresponds to a point on the line connecting the fraction with 0,0. Consider the point 6/9 ($y = 6$, $x = 9$). It is not visible from 0,0 because it can be simplified to 2/3. Place a straightedge so that it joins 0,0 and 6/9 and you will see that the visibility of 6/9 is blocked by the point at

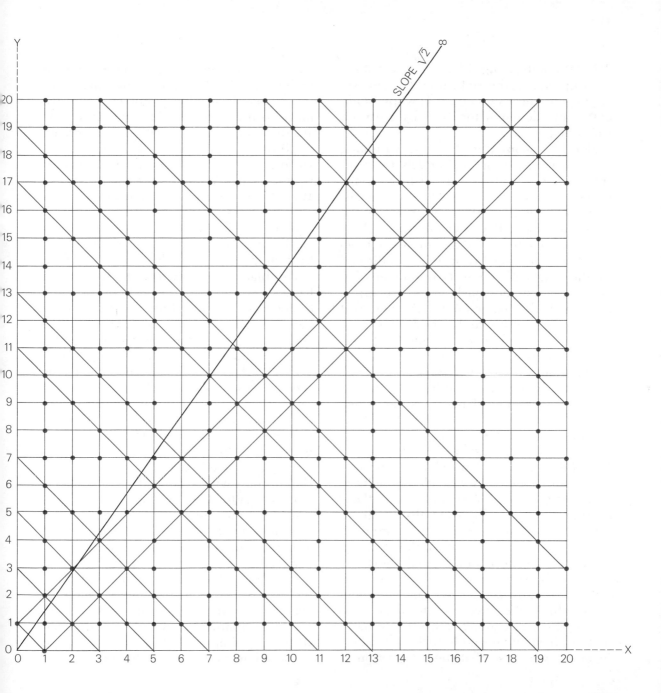

SLOPE √2̅

∞

2/3. All points along the diagonals that extend up and to the right from 0/1 and 1/0 are visible because no fraction whose numerator and denominator differ by 1 can be simplified.

Note that many of the diagonals running the other way—from upper left to lower right—consist entirely of visible points except for their ends. All these diagonals, Ely points out, cut the coordinate axes at prime numbers. Every visible point along such a diagonal is a fraction formed by two numbers that sum to the prime indicated by the diagonal's ends. Two numbers that sum to a prime obviously must be coprime (if they had a common factor, then that factor would also evenly divide the sum), so such fractions cannot be simplified. Vertical and horizontal lines that cut an axis at a prime get progressively denser with visible lattice spots as the primes get larger, because such lines have invisible lattice points only where the other coordinate is a multiple of the prime.

Is it possible to stand at 0,0 and look into this infinite orchard along a line that will never, even when extended to infinity, intersect a "tree"? Yes; not only is there an infinity of such lines but also there are infinitely more of them than there are lines that hit trees! Consequently if the direction for a line of sight is determined randomly, the probability of finding a tree along that line is virtually zero. How can we define such a line? We have only to slope it so that every point along it has coordinates that are incommensurable with each other; in other words, so that the y/x fraction of any point—which is the same as the tangent of the angle that the sloping line makes with the x axis—is irrational. For example, we move to the right along the x axis to, say, 10, then up to a point with a y coordinate of 10 times pi. If we join this point to 0,0, we produce a line that cannot, no matter how far it is extended, hit a point because 10pi/10 equals pi, an irrational number. (It would take some fine drawing and a superpowered microscope to detect how far the line misses the point at 355/113. This fraction gives pi to six decimal places!)

The black line shown in the illustration has a slope of $\sqrt{2}$. It is easy to prove that a bullet traveling this line could not, from here to eternity, strike a tree. The right triangle shown in Figure 148 has a base of 1 and an altitude of $\sqrt{2}$, so the tangent of angle θ is $\sqrt{2}$. If we extend the hypotenuse as shown by the broken line to form any larger right triangle on the extended base line, the altitude and base of the larger triangle will have the same irrational ratio. The bases and altitudes of all such triangles correspond to the two coordinates of the sloping line with a tangent of $\sqrt{2}$. Therefore, no matter how far the sloping line is extended into the lattice of integers, the coordinates of any point along that line will form the same irrational fraction. But every lattice point represents a *rational* fraction; therefore no lattice point can be on the line.

Observe, however, that by searching for near misses we can find fractions that are excellent approximations of the irrational slope. Think of the $\sqrt{2}$ line as a taut rope anchored at infinity. If we hold the end at

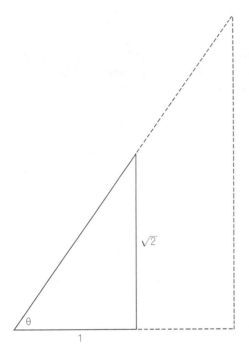

$\sqrt{2}$

θ

1

148. *Tangent of angle θ equals* $\sqrt{2}$

One of the simplest ways to express $\sqrt{2}$ is by the endless continued fraction

$$1 + \frac{1}{2 + \cfrac{1}{2 + \cfrac{1}{2 + 1}}} \cdots$$

If we start at the top and form partial sums (that is, 1, 1 + 1/2, 1 + 1/3 and so on), we get just those fractions mentioned above: 1, 3/2, 4/3, 7/5, 10/7, 17/12. They come closer and closer to $\sqrt{2}$ as their lattice points come closer and closer to the sloping line.

The discussion of irrational numbers suggests the following problem: Let the coordinates of a point be $y = \sqrt{27}$, $x = \sqrt{3}$. Does the infinite line passing from the origin through this point cut any points other than 0,0?

If a billiard ball is placed at 0,0 and stroked so that it travels up the main diagonal at an angle of 45 degrees, it will of course continue forever, passing only through lattice points whose fractions reduce to 1 (the tangent of 45 degrees). Now suppose we confine the lattice to rectangles of arbitrary size, provided that heights and widths are integral, and assume that the ball rebounds from all sides and rolls without friction over the surface of the latticed billiard table. It is not hard to show, by a reflection technique depicted in Hugo Steinhaus' *Mathematical Snapshots*, that whatever the dimensions of the rectangle, the ball will strike one of the table's

0,0 and move the rope east, it will press against trees that represent fractions smaller than $\sqrt{2}$, or 1.4142136. . . , but that get closer to $\sqrt{2}$ as one moves away from 0,0. The first tree it touches is 1/1, or 1, a poor approximation. The next is 4/3, a bit better, and the next is 7/5, or 1.4, which is not bad. Similarly, if we move the end of the rope northward, it presses against fractions larger than 1.4142136. . . , but the excess approaches 0 as we move toward infinity. The first fraction, 2/1, or 2, is not very good; 3/2 is better, 10/7 still better and 17/12, or 1.41666. . . , misses $\sqrt{2}$ by only .0024+.

other three corners after a finite number of bounces.

This statement can be made stronger. Regardless of the angle of the first shot, if the ball strikes the first cushion at a point that is a rational distance from a corner, it will eventually strike one of the table's corners. But if it hits the first cushion at an irrational spot, every rebound will be at an angle with an irrational tangent and the path will never touch a lattice point. Since the corners are lattice points, the ball will never strike a corner. There are infinitely more irrational points on a line than there are rational points. Therefore the probability is infinitely close to zero that an ideal ball (we must think of the ball as a point) shot from the corner at a random angle will strike the first cushion at a rational point. Imagine the table covered with a fine screen of lattice points — billions of them, all with rational coordinates. The randomly shot ball will move forever around the table, never going over a path twice, never once touching a single lattice point.

Here we are concerned only with the simpler case of a ball traveling along diagonals that form 45-degree angles with the table's sides. An intriguing question (first sent to me by Joseph Becker of Milwaukee) immediately arises. Given the table's dimensions, how can one predict which of the three corners the ball will hit? We can always draw a graph and find out, but if the table has, say, a width of 10,175 units and a length of 11,303 units, graphing a solution would be tedious.

As Becker points out, if at least one side of the table is odd, a clever parity check leads to simple rules for determining which corner the ball will hit. Suppose both sides are odd. We color the 0,0 point and every second lattice point [see "a" in Figure 149]. Clearly the ball will pass through the colored points only. Of the three possible terminating corners, only the northeast corner is colored, so this must be the corner the ball will strike. (The reader can verify this by continuing the ball's path through the colored points.) If one side of the table is even and the other odd, the same parity coloring ["b" and "c"] shows that the ball must strike the corner adjacent to the origin and on the table's even side.

When both sides of the table are even, we run into an unforeseen difficulty. There are colored spots on *all four* corners [d]. Which of the three possible terminal corners will the ball hit? A little experimenting on graph paper will show that all three can be reached on various even-even tables. Can the reader devise an arithmetical rule for quickly determining, on any table with even sides, which corner the ball will hit?

A hint for the solution to this problem lies in the curious fact that the point on the table's longest side that is nearest the origin, and on the ball's path as well, is always exactly twice the greatest common divisor (gcd) of the two sides. If the two sides are coprime, then of course the gcd is 1. This is the case in *a* and *c* of Figure 149. Sure enough, on the longest side we see that the point on the ball's path nearest to 0,0 is 2, or twice the gcd.

This property of 45-degree paths of a

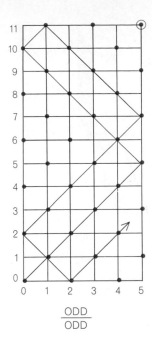

a

$$\frac{\text{ODD}}{\text{ODD}}$$

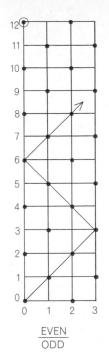

b

$$\frac{\text{EVEN}}{\text{ODD}}$$

149. Parity coloring checks for billiard-ball paths

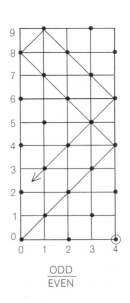

c

$$\frac{\text{ODD}}{\text{EVEN}}$$

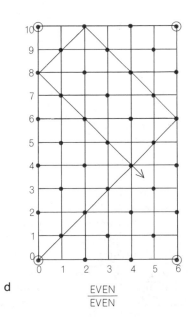

d

$$\frac{\text{EVEN}}{\text{EVEN}}$$

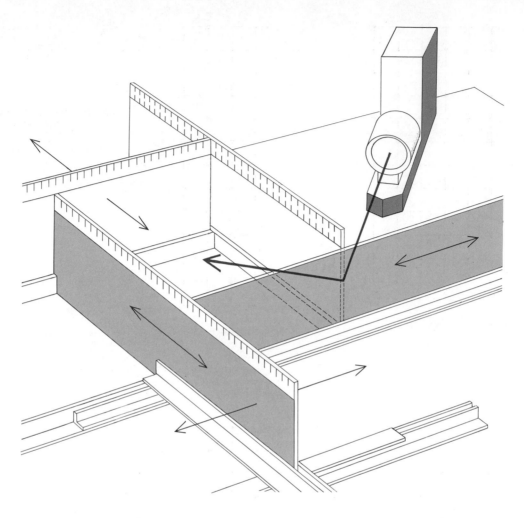

150. *Zavrotsky's device for finding greatest common divisors*

bouncing ball inside a rectangular lattice of integers suggested to Andrés Zavrotsky, of the University of the Andes in Venezuela, an optical device for finding the greatest common divisors of pairs of integers. A sketch of his invention (U.S. patent 2,978,-816, April 11, 1961) is shown in Figure 150. Four mirrors with integral scales on their edges can be adjusted to form a rectangle with sides equal to the pair of numbers under investigation. A pencil of light is introduced through a crack at one corner, as shown. It rebounds at an angle of 45 degrees from the corner—the zero point on the two scales meeting at that corner—and continues its path from mirror to mirror until it terminates at one of the other three corners. The illuminated mark closest to the corner of origin of what Zavrotsky calls the "optical billiard" on the rectangle's long side is twice the gcd. Zavrotsky intended his invention to serve as a teaching device. Readers should have little difficulty proving that the device cannot fail to work, and solving this problem: Given the rectangle's sides, find a formula for the total length of the light's path, from 0,0 to the corner; also find a formula for the number of times the "optical billiard" rebounds from a side.

By connecting pairs of lattice points with straight lines one can draw an infinite variety of simple polygons [*see Figure 151*]. ("Simple" here means that no side crosses another.) The area of such a "lattice polygon" can be calculated by the tiresome method of cutting it up into simpler figures, but again there is an easier and more amusing way to do it. We apply the following remarkable theorem: The area of any lattice polygon is one-half the number of lattice points on its border, plus the number of points inside its border, minus one. The unit of area is the area of the "unit cell" of the lattice.

This beautiful theorem, which Steinhaus says was first published by one G. Pick in a Czechoslovakian journal in 1899, belongs to "affine" geometry, a geometry that plays an important role in the mathematics of relativity. This means that the theorem holds even when the lattice is distorted by stretching and shearing. For example, the formula applies to the connect-the-dot polygon on the lattice shown in Figure 152. As before,

the unit area is the unit cell, in this case the little parallelogram to the right. This *T*-polygon, like its counterpart above, has 24 points on its border and 9 inside; according to Pick's formula, its area is $12 + 9 - 1 = 20$ unit cells, as is easily verified. Readers may enjoy seeing if they can devise a complete proof of the theorem. An outline of one such proof is given in H. S. M. Coxeter's *Introduction to Geometry* (New York: John Wiley and Sons, 1961; page 209).

One is tempted to suppose that it would be easy to extend Pick's formula to polyhedrons drawn on integral lattices in three dimensions. Figure 153 quickly dispels this illusion. It shows the unit cell at the 0,0,0 corner of a three-space cubical coordinate system. The four points at 0,0,0, 1,0,0, 0,1,0, and 1,1,1 mark the corners of a lattice tetrahedron. If we raise the apex of this pyramid to 1,1,2, we increase the tetrahedron's volume but no new lattice points appear on its edges or faces or in the interior. Indeed, by raising the apex higher along the same coordinate the volume can be made as large as we please without increasing the number of lattice points involved. It is possible, however, to find a formula by introducing a secondary lattice. The interested reader will find this explained in "On the Volume of Lattice Polyhedra," by J. E. Reeve, in *Proceedings of the London Mathematical Society;* July, 1957, pages 378–395. For an extension of the formula to still higher spaces, see "The Volume of a Lattice Polyhedron," by I. G. Macdonald, in *Proceedings of the Cambridge Philosophical Society;* October, 1963; pages 719–726.

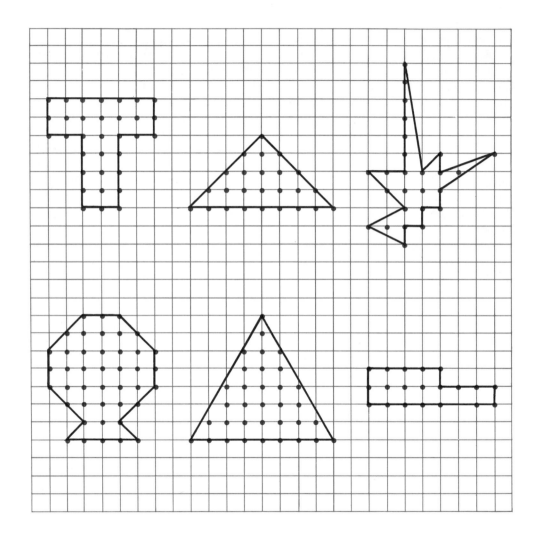

151. *Find the area of these "lattice polygons"*

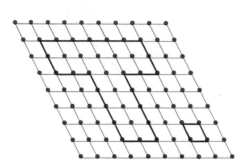

152. "Affine" transformation of lattice polygon

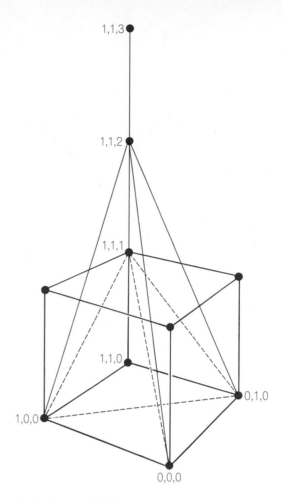

153. Lattice tetrahedrons

A final problem: On the square lattice of integers, connect exactly 12 lattice points to form a lattice polygon of the same shape as the *T*-polygon in Figure 152 but with an area of ten square units. (According to Pick's formula, it must surround exactly five lattice points.)

Answers

A line from 0,0 on the lattice of integers, with a slope of $\sqrt{27}/\sqrt{3}$, will pass through an infinity of lattice points. Because $\sqrt{27} = \sqrt{3 \times 9} = 3\sqrt{3}$, the fraction $\sqrt{27}/\sqrt{3}$ reduces to 3/1, a rational fraction. The first lattice point on this slope is $y = 3$, $x = 1$.

On a rectangular lattice with even sides, a ball leaving the origin at a 45-degree angle will travel through lattice points separated (along coordinate lines) by a distance equal to twice the greatest common divisor (gcd) of the sides. If we mark these points with spots as in Figure 154, we see

that only one of the three possible terminal corners receives a spot, and it therefore must be the corner the ball will hit. To determine which corner this will be, we divide each side by the gcd. If both results are odd, the ball strikes the corner diagonally opposite the origin. If one result is even (both cannot be even), the ball strikes the corner on that side and adjacent to the

217

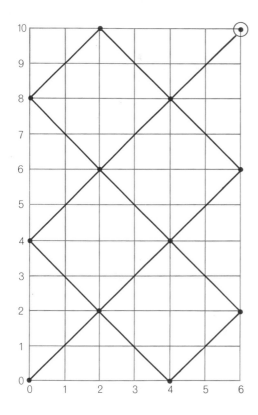

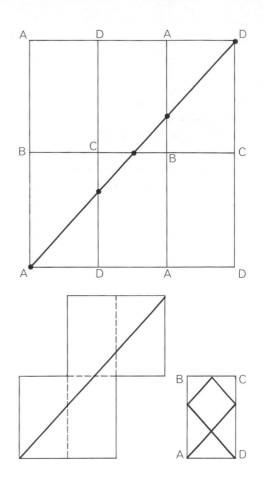

154. *Solution to the "even-even" problem*

155. *Finding the length of the ball's path*

origin. For rules governing the more general case, when the ball's path may be at any angle with a rational slope, see M. S. Klamkin's solution to his problem No. 116 in the *Pi Mu Epsilon Journal*, Spring, 1963.

Formulas for the length of the ball's path and the number of rebounds are intuitively evident in Figure 155, adapted from Hugo Steinhaus' *Mathematical Snapshots*. What-

ever the integral dimensions of a rectangle, a square can always be formed by placing a finite number of replicas of the rectangle side by side as shown at the top in the illustration. The smallest square formed in this way will have a side that is the lowest common multiple of the rectangle's two sides.

Think of each rectangle as a mirror reflection of each rectangle adjacent to it. The diagonal line from *A*, where the ball starts its 45-degree path, to the opposite corner will then be the "unfolded" path, so to speak, of the ball as it rebounds from side to side. If we cut out just those rectangles that contain the path *(lower left)*, fold them along the broken lines and then hold the packet up to a strong light, the diagonal line will trace the actual path of the ball around the rectangle *(lower right)*.

Since the diagonal line *AD*, on the large square, is the hypotenuse of a right isosceles triangle with a side equal to the lowest common multiple of the sides of the rec-

tangle, we see at once that the length of the path is this lowest common multiple times $\sqrt{2}$.

The spots shown along the diagonal, minus the end spots, represent points of rebound. It is easy to see that the number of rebounds must be

$$\frac{a+b}{\gcd} - 2 \, ,$$

where *a* and *b* are the sides of the original rectangle and gcd is their greatest common divisor.

Figure 156 shows the only way to draw the *T*-polygon on a square lattice so that there are 12 points on the border and five inside: an area of ten square units.

156. The T-polygon solution

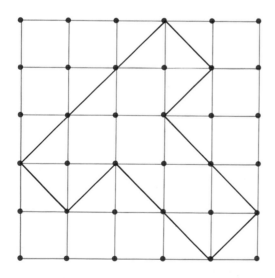

References

"Areas of Simple Polygons." David A. Kelley. *The Pentagon*, Vol. 20, No. 1; Fall, 1960. Pages 3–11.

"Lattice Paths with Diagonal Steps." L. Moser and W. Zayachkowski. *Scripta Mathematica*, Vol. 26, No. 3; Autumn, 1963. Pages 223–229.

"Lattice Points and Polygonal Area." Ivan Niven and H. S. Zuckerman. *The American Mathematical Monthly*, Vol. 74, No. 10; December, 1967. Pages 1195–1200.

Mathematical Snapshots. Hugo Steinhaus. 3rd ed. New York: Oxford University Press, 1968.

22. Infinite Regress

THE INFINITE REGRESS, along which thought is compelled to march backward in a never ending chain of identical steps, has always aroused mixed emotions. Witness the varied reactions of critics to the central symbol of Broadway's most talked-about 1964 play, Edward Albee's *Tiny Alice*. The principal stage setting—the library of an enormous castle owned by Alice, the world's richest woman—is dominated by a scale model of the castle. Inside it lives Tiny Alice. When lights go on and off in the large castle, corresponding lights go on and off in the small one. A fire erupts simultaneously in castle and model. Within the model is a smaller model in which a tinier Alice perhaps lives, and so on down, like a set of nested Chinese boxes. ("Hell to clean," comments the butler, whose name is Butler.) Is the castle itself, into which the play's audience peers, a model in a still larger model, and that in turn . . .? A similar infinite nesting is the basis of E. Nesbit's short story, "The Town in the Library in the Town in the Library" (in her *Nine Unlikely Tales*); perhaps this was the source of Albee's idea.

For many of the play's spectators the endless regress of castles stirs up feelings of anxiety and despair: Existence is a mysterious, impenetrable, ultimately meaningless labyrinth; the regress is an endless corridor that leads nowhere. For theological students, who are said to be flocking to the play, the regress deepens an awareness of what Rudolf Otto, the German theologian, called the *mysterium tremendum:* the ultimate mystery, which one must approach with awe, fascination, humility and a sense of "creaturehood." For the mathematician and the logician the regress has lost most of its terrors; indeed, as we shall soon see, it is a powerful, practical tool even in recreational mathematics. First, however, let us glance at some of the roles it has played in Western thought and letters.

Aristotle, taking a cue from Plato's *Parmenides,* used the regress in his famous "third man" criticism of Plato's doctrine of ideas. If all men are alike because they have something in common with Man, the ideal and eternal archetype, how (asked Aristotle) can we explain the fact that one man and Man are alike without assuming another archetype? And will not the same reasoning demand a third, fourth, and fifth archetype, and so on into the regress of more and more ideal worlds?

A similar aversion to the infinite regress underlies Aristotle's argument, elaborated by hundreds of later philosophers, that the cosmos must have a first cause. William Paley, an eighteenth-century English theologian, put it this way: "A chain composed of an infinite number of links can no more support itself than a chain composed of a finite number of links." A finite chain does indeed require support, mathematicians were quick to point out, but in an infinite chain *every* link hangs securely on the one above. The question of what supports the entire series no more arises than the question of what kind of number precedes the infinite regress of negative integers.

Agrippa, an ancient Greek skeptic, argued that nothing can be proved, even in mathematics, because every proof must be proved valid and its proof must in turn be proved, and so on. The argument is repeated by Lewis Carroll in his paper "What the Tortoise Said to Achilles" (*Mind,* April, 1895). After finishing their famous race, which involved an infinite regress of smaller and smaller distances, the Tortoise traps his fellow athlete in a more disturbing regress. He refuses to accept a simple deduction involving a triangle until Achilles has written down an infinite series of hypothetical assumptions, each necessary to make the preceding argument valid.

F. H. Bradley, the English idealist, argued (not very convincingly) that our mind cannot grasp *any* type of logical relation. We cannot say, for example, that castle A is smaller than castle B and leave it at that, because "smaller than" is a relation to which both castles are related. Call these new relations c and d. Now we have to relate c and d to the two castles and to "smaller than." This demands four more relations, they in turn call for eight more, and so on, until the shaken reader collapses into the arms of Bradley's Absolute.

In recent philosophy the two most revolutionary uses of the regress have been made by the mathematicians Alfred Tarski and Kurt Gödel. Tarski avoids certain troublesome paradoxes in semantics by defining truth in terms of an endless regress of "metalanguages," each capable of discussing the truth and falsity of statements on the next lower level but not on its own level. As Bertrand Russell once explained it: "The man who says 'I am telling a lie of order n' is telling a lie, but a lie of order $n + 1$." In a closely related argument Gödel was able to show that there is no single, all-inclusive mathematics but only an infinite regress of richer and richer systems.

The endless hierarchy of gods implied by so many mythologies and by the child's inevitable question "Who made God?" has appealed to many thinkers. William James closed his *Varieties of Religious Experience* by suggesting that existence includes a collection of many gods, of different degrees of inclusiveness, "with no absolute unity realized in it at all. Thus would a sort of polytheism return upon us. . . ." The notion turns up in unlikely places. Benjamin Franklin, in a quaint little work called *Articles of Belief and Acts of Religion,* wrote: "For I believe that man is not the most perfect being but one, but rather that there are many degrees of beings superior to him." Our prayers, said Franklin, should be directed only to the god of our solar system, the deity closest to us. Many writers have viewed life as a board game in which we are the pieces moved by higher intelligences who in turn are the pieces in a vaster game. The prophet in Lord Dunsany's story "The South Wind" observes the gods striding through the stars, but as he worships them he sees the outstretched hand of a player "enormous over Their heads."

Graphic artists have long enjoyed the infinite regress. Who can look at the striking cover of the April, 1965, issue of *Scientific American* (showing the magazine cover reflected in the pupil of an eye) without recalling, from his childhood, a cereal box or magazine cover on which a similar trick was played? The cover of the November, 1964, *Punch* showed a magician pulling a rabbit out of a hat. The rabbit in turn is pulling a smaller rabbit out of a smaller hat, and this endless series of rabbits and hats moves up and off the edge of the page. It is not a bad picture of contemporary particle physics. The latest theory proposes a smaller, yet undetected, group of particles called "quarks" to explain the structure of known particles. Is the cosmos itself a particle in some unthinkably vast variety of matter? Are the laws of physics an endless regress of hat tricks?

The play within the play, the puppet show within the puppet show, the story within the story have amused countless writers. Luigi Pirandello's *Six Characters in Search of an Author* is perhaps the best-known stage example. The protagonist in Miguel de Unamuno's novel *Mist,* anticipating his death later in the plot, visits Unamuno to protest and troubles the author with the thought that he too is only the figment of a higher imagination. Philip Quarles, in Aldous Huxley's *Point Counter Point,* is writing a novel suspiciously like *Point Counter Point.* Edouard, in André Gide's *The Counterfeiters*, is writing *The Counterfeiters.* Norman Mailer's story "The Notebook" tells of an argument between the writer and his girl friend. As they argue he jots in his notebook an idea for a story that has just come to him. It is, of course, a story about a writer who is arguing with his girl friend when he gets an idea. . . .

J. E. Littlewood, in *A Mathematician's Apology,* recalls the following entry, which won a newspaper prize in Britain for the best piece on the topic: "What would you most like to read on opening the morning paper?"

OUR SECOND COMPETITION

The First Prize in the second of this year's competitions goes to Mr. Arthur Robinson, whose witty entry was easily the best of those we received. His choice of what he would like to read on opening his paper was headed "Our Second Competition" and was as follows: "The First Prize in the second of this year's competitions goes to Mr. Arthur Robinson, whose witty entry was easily the best of those we received. His choice of what he would like to read on opening his paper was headed 'Our Second Competition,' but owing to paper restrictions we cannot print all of it."

One way to escape the torturing implications of the endless regress is by the topological trick of joining the two ends to make a circle, not necessarily vicious, like the circle of weary soldiers who rest themselves in a bog by each sitting on the lap of the man behind. Albert Einstein did exactly this when he tried to abolish the endless regress of distance by bending three-dimensional space around to form the hypersurface of a four-dimensional sphere. One can do the same thing with time. There are Eastern religions that view history as an endless recurrence of the same events. In the purest sense one does not even think of cycles following one another, because there is no outside time by which the cycles can be counted; the *same* cycle, the *same* time go around and around. In a similar vein, there is a sketch by the Dutch artist Maurits C. Escher of two hands, each holding a pencil and sketching the other [*see Figure 157*]. In *Through the Looking Glass* Alice dreams of the Red King, but the King is himself asleep and, as Tweedledee points out, Alice is only a "sort of thing" in *his* dream. *Finnegans Wake* ends in the middle of a sentence that carries the reader back for its completion to the broken sentence that opens the book.

Since Fitz-James O'Brien wrote his pioneer yarn "The Diamond Lens" in 1858 almost countless writers have played with the theme of an infinite regress of worlds on smaller and smaller particles. In Henry Hasse's story "He Who Shrank" a man on a cosmic level much larger than ours is the victim of a scientific experiment that has caused him to shrink. After diminishing through hundreds of subuniverses he lingers just long enough in Cleveland to tell his story before he vanishes again, wondering how long this will go on, hoping that the levels are joined at their ends so that he can get back to his original cosmos.

Even the infinite hierarchy of gods has been bent into a closed curve by Dunsany in his wonderful tale "The Sorrow of Search." One night as the prophet Shaun is observing by starlight the four mountain gods of old—Asgool, Trodath, Skun, and

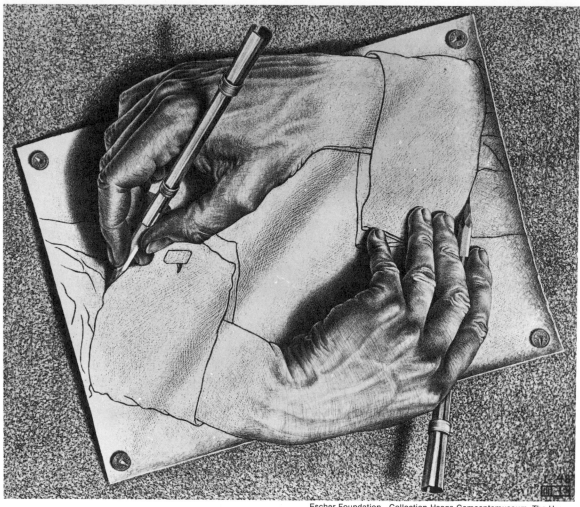

Escher Foundation—Collection Haags Gemeentemuseum, The Hague

157. Maurits C. Escher's "Drawing Hands"

Rhoog—he sees the shadowy forms of three larger gods farther up the slope. He leads his disciples up the mountain only to observe, years later, two larger gods seated at the summit, from which they point and mock at the gods below. Shaun takes his followers still higher. Then one night he perceives across the plain an enormous, solitary god looking angrily toward the mountain. Down the mountain and across the plain goes Shaun. While he is carving on rock the story of how his search has ended at last with the discovery of the ultimate god, he sees in the far distance the dim forms of four higher deities. As the reader can guess, they are Asgool, Trodath, Skun, and Rhoog.

No branch of mathematics is immune to the infinite regress. Numbers on both sides of zero gallop off to infinity. In modular arithmetics they go around and around. Every infinite series is an infinite regress. The regress underlies the technique of mathematical induction. Georg Cantor's transfinite numbers form an endless hierarchy of richer infinities. A beautiful modern example of how the regress enters into a mathematical proof is related to the difficult problem of dividing a square into other squares no two of which are alike (see Chapter 17 of my *Second Scientific American Book of Mathematical Puzzles and Diversions*; New York: Simon and Schuster, 1965). The question arises: Is it possible similarly to cut a cube into a finite number of smaller cubes no two of which are alike? Were it not for the deductive power of the regress, mathematicians might still be searching in vain for ways to do this. The proof of impossibility follows.

Assume that it is possible to "cube the cube." The bottom face of such a dissected cube, as it rests on a table, will necessarily be a "squared square." Consider the smallest square in this pattern. It cannot be a corner square, because a larger square on one side keeps any larger square from bordering the other side [*see "a" in Figure 158*]. Similarly, the smallest square cannot be elsewhere on the border, between corners, because larger squares on two sides prevent a third larger square from touching the third side [*b*]. The smallest square must therefore be somewhere in the pattern's interior. This in turn requires that the smallest cube touching the table must be surrounded by cubes larger than itself. This is possible [*c*], but it means that four walls must rise above all four sides of the small cube — preventing a larger cube from resting on top of it. Therefore on this smallest cube there must rest a set of smaller cubes, the bottoms of which will form another pattern of squares.

The same argument is now repeated. In the new pattern of squares the smallest square must be somewhere in the interior. On this smallest square must rest the smallest cube, and the little cubes on top of it will form another pattern of squares. Clearly the argument leads to an endless regress of smaller cubes, like the endless hierarchy of fleas in Dean Swift's jingle. This contradicts the original assumption that the problem is solvable.

Geometric constructions such as this one, involving an infinite regress of smaller figures, sometimes lead to startling results. Can a closed curve of infinite length enclose a finite area of, say, one square inch? Such pathological curves are infinite in number. Start with an equilateral triangle [*see "a" in Figure 159*] and on the central third of each side erect a smaller equilateral triangle. Erase the base lines and you have a six-pointed star [*b*]. Repeating the construction on each of the star's 12 sides produces a 48-sided polygon [*c*]. The third step is shown in *d*. The limit of this infinite construction, called the snowflake curve, bounds an area 8/5 that of the original triangle. It is easy to show that successive

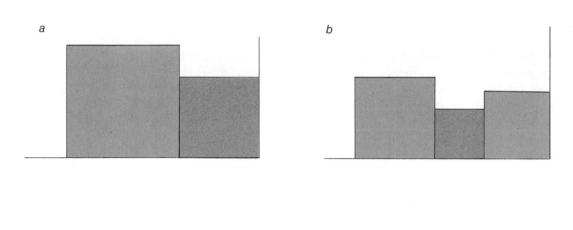

a

b

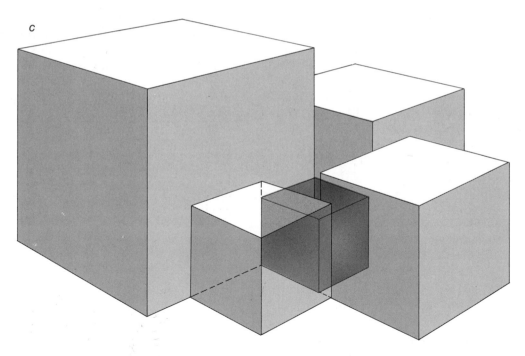

c

158. Proof that the cube cannot be ''cubed''

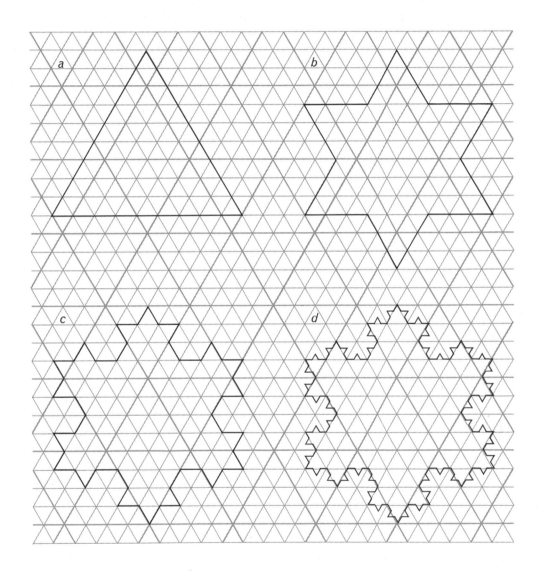

159. *The snowflake curve*

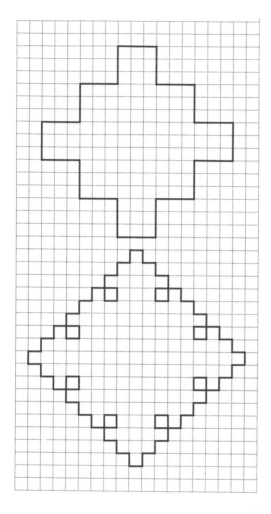

160. The cross-stitch curve

additions of length form an infinite series that diverges; in short, the length of the snowflake's perimeter is infinite. (In 1956 W. Grey Walter, the British physiologist, published a science-fiction novel, *The Curve of the Snowflake*, in which a solid

analogue of this crazy curve provides the basis for a timetravel machine.)

Here are two easy puzzles about the less well known square version of the snowflake, a curve that has been called the cross-stitch. On the middle third of each side of a unit square erect four smaller squares as shown at the top of Figure 160. The second step is shown at the bottom. (The squares will never overlap, but corners will touch.) If this procedure continues to infinity, how long is the final perimeter? How large an area does it enclose?

Answers

The cross-stitch curve has, like its analogue the snowflake, an infinite length. It bounds an area twice that of the original square. The drawing at the left in Figure 161 shows its appearance after the third construction. After many more steps it resembles (when viewed at a distance) the drawing at the right. Although the stitches seem to run diagonally, actually every line segment in the figure is vertical or horizontal. Similar constructions of pathological curves can be based on any regular polyhedron, but beyond the square the figure is muddied by overlapping, so that certain conventions must be adopted in defining what is meant by the enclosed area.

Samuel P. King, Jr., of Honolulu, supplied a good analysis of curves of this type, including a variant of the cross-stitch discovered by his father. Instead of erecting four squares outwardly each time, they are

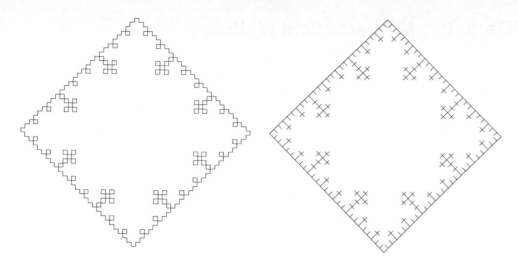

161. Solution to cross-stitch curve problem

erected inwardly from the sides of each square. The limit curve has an infinite length, but encloses zero area.

References

"The Infinite Regress." John Passmore. In his *Philosophical Reasoning*. New York: Charles Scribner's Sons, 1961.

"Snowflake Curves." Bruce W. King. *The Mathematics Teacher,* Vol. 57, No. 4; April, 1964. Pages 219–222.

"Avatars of the Tortoise." Jorge Luis Borges. In his *Labyrinths: Selected Stories and Other Writings.* New York: New Directions, 1964.

"Partial Magic in the Quixote." Jorge Luis Borges. In his *Labyrinths: Selected Stories and Other Writings.* New York: New Directions, 1964.

"A Generalization of the Von Koch Curve." Joel E. Schneider. *Mathematics Magazine,* Vol. 38, No. 3; May, 1965. Pages 144–147.

"Infinity in Mathematics and Logic." James Thomson. In *The Encyclopedia of Philosophy.* Vol. 4. New York: Crowell Collier, 1967. Pages 183–190.

"Infinity in Theology and Metaphysics." H. P. Owen. In *The Encyclopedia of Philosophy.* Vol. 4. New York: Crowell Collier, 1967. Pages 190–193.

23. O'Gara, the Mathematical Mailman

I'm seen in the country, I'm seen
  in the town,
I'm servant of all from pauper
  to crown.
Take one letter from me, and still
  my good name
In spite of your action continues
  the same.
Take from me two letters, then three
  and then four,
My name will continue the same
  as before.
In fact, you can take all the letters
  I've got
And my name you will not have altered
  one jot!

I WAS LEAFING through a stack of unopened envelopes on my desk, looking for unusual foreign postage stamps, when a bright red sticker caught my eye. It said: "Please notify P. O. immediately if this gummed label has fallen off in transit."

This had such an unmistakable flavor of an "Irish bull" (the essence of which is logical contradiction) that I was not surprised to discover that "P. O." stood not for "Post Office" but for the sender, one Patrick O'Gara of Brooklyn. His letter began with the charade above, which he said a grandfather in Ireland had clipped from an English newspaper half a century ago. O'Gara was a postman by profession but a recreational mathematician by avocation. He entertained himself on his daily rounds, he said, by creating unusual puzzles. Would I be interested in discussing some of them with him?

The intersection set of all people interviewed in this book and the set of all existing individuals is, I must confess, empty. This, however, has never discouraged me from further interviews. Moreover, I was psychologically prepared for meeting a remarkable mailman, having recently reread one of my favorite Father Brown stories, "The Invisible Man." In this G. K. Chesterton murder mystery four witnesses swear that no one has entered or left a certain building because they all take the postman so much for granted that they do not consider him worth mentioning, "Nobody ever notices postmen, somehow," as Father Brown put it, "yet they have passions like other men. . . ."

Although O'Gara's major passion was recreational mathematics, he minored in philately. He turned out to be a young, athletically built chap with sandy hair and a face heavily freckled by the sun. His education had not gone beyond high school, but the small study in his bachelor apartment in Brooklyn Heights overflowed with old and new books on mathematical puzzles, and after a few minutes of conversation it was obvious that he was well informed in the field.

"Are you a stamp collector?" he asked.

"No," I replied, "but my ten-year-old son has just started an album."

"Encourage him to specialize," said O'Gara. "The big thing now, you know, is what is called thematic or topical collecting. Nobody collects miscellaneously anymore. Let me show you some of my topicals."

His largest collection concerned mathematics. I was amazed by the number of eminent mathematicians whose portraits had appeared on these little engravings ever since Germany, in 1926, had issued the first mathematical stamp: a 40-pfennig violet with the head of Leibniz. O'Gara had French stamps honoring Descartes, Pascal, Buffon, Carnot, Laplace, Poincaré and many others; Italian stamps showing scenes from the life of Galileo; Dutch stamps with faces of Huygens, Lorentz and others; Russian stamps honoring such notables as Euler, Chebyshëv and Lobachevski. A striking set of four Norwegian stamps commemorated the centenary in 1929 of Abel's death. Two stamps issued by the Irish Free State in 1943 bore portraits of Hamilton to celebrate the centenary of his discovery of quaternions. Gauss appeared on a German stamp in 1955. A Romanian mathematics journal, *Gazeta Matematica,* was honored on its 50th birthday with a pair of stamps, and in 1955 Greece commemorated the 2,500th anniversary of the Pythagorean school by putting a 3–4–5 right triangle on four stamps [*see Figure 162*]. A French stamp honoring Descartes in 1937 is of special interest because the first issue showed an incorrect title of his greatest work. (The title was corrected on the second issue.)

"Has the United States ever honored a mathematician with a commemorative stamp?" I asked.

O'Gara shook his head. "Neither has England, but of course England has an excuse. She limits her stamp portraits to members of the royal family." (In 1966 a U.S. 8-cent purple bore Einstein's picture, but Einstein was not primarily a mathematician.)

One of O'Gara's most amusing topical collections contained what he called "science goofers"—stamps on which someone

162. A 1955 Pythagorean postage stamp

163. Serbian and U.S. stamps that reveal hidden pictures when turned upside down

had made a whopping scientific mistake. The British colony Saint Kitts-Nevis issued a stamp in 1903 showing Columbus, on deck, searching for land with a telescope, which had not yet been invented. A skier's ears, on a 1934 Austrian stamp, are upside down. The constellation of the Southern Cross somehow got reversed when it appeared on a 1940 Brazilian stamp. A U.S. Transcontinental Railroad commemorative of 1944 shows smoke from a locomotive billowing to one side and a flag blowing the other way.

Another unusual thematic collection consisted of "hidden pictures." In 1904 Serbia issued a famous "death mask" stamp: the profiles of Karageorge and Peter I Karageorgevich, upside down, merge to form a death mask of the Serbian king Alexander I Obrenovich—who had been murdered the year before by Karageorgevich's supporters. On a 1932 U.S. three-cent, the tie and shirtfront of Daniel Webster turn into the face of Fu Manchu when viewed upside down

[see Figure 163]. A range of mountains on a 1934 U.S. National Parks issue becomes a man's profile when rotated 90 degrees. A 1935 Boulder Dam stamp, inverted, looks like the Liberty Bell. On a West German 50-pfennig of 1964 a tiny face of Hitler is concealed in some tree foliage. O'Gara had scores of others.

When a postman goes on a holiday, the fictional detective Charlie Chan used to say, he takes long walks. During a vacation in Europe a few years ago O'Gara had actually made, he told me, a special trip to the Baltic seaport of Kaliningrad (formerly Königsberg, the capital of East Prussia) for the sole purpose of tramping over the famous seven bridges of Königsberg in one continuous path without going over any bridge

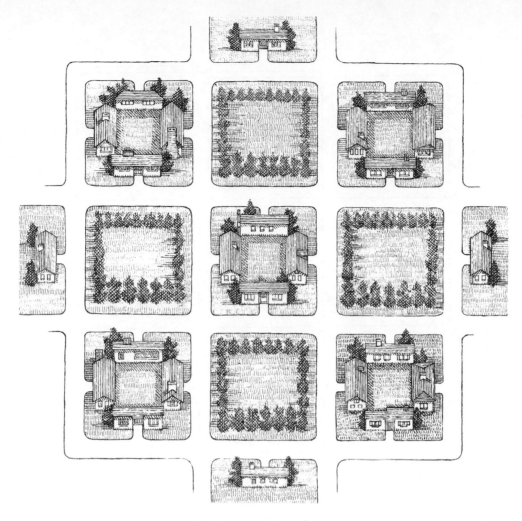

164. What is the best route for delivering to each home?

twice. He was able to do this, he explained, because an eighth bridge had been built across the Pregel River since Leonhard Euler first proved that the original problem was unsolvable. On a day off last winter, fortified by some Irish whisky, O'Gara had conducted extensive investigations of random-walk problems in a large open field of snow somewhere in Brooklyn.

"I'm very good at visualizing geometric patterns," he told me. "Used to play a lot of blindfold chess as a boy. So I work on my graph puzzles in my head while I'm making my rounds. For instance . . ."

He paused to sketch for me an aerial view of a housing development where he had at one time delivered mail [*see Figure 164*]. There were houses in every second block,

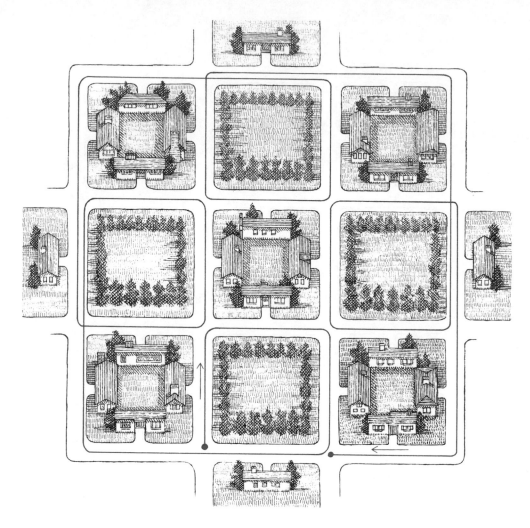

165. *A minimum-length route with right turns only*

and each house required a delivery, as shown on the map. "It's easy to apply Euler's rules here," said O'Gara. "They show that it's not possible to make mail deliveries along all eight streets without walking some of the blocks more than once." (To trace a network in an unbroken path, without going over any part twice, there must be either *no* intersections where an odd num-ber of paths meet or *exactly two* such intersections.) "But how *short* can the path be? I soon convinced myself it couldn't be less than 27 blocks. Every day for months I tried to find new 27-block paths that would meet various restraints. For example, I found all sorts of ways to cover the eight streets in a 27-block path without ever making a left turn [*see Figure 165*]. Finally I hit on two pretty problems that I believe your readers might like."

The first problem, O'Gara explained, is to find a path that covers all the streets in the minimal length of 27 blocks and that also has the minimal number of turns. The path in the illustration, for instance, has 19 turns —far more than necessary. A "turn" occurs at any point where the path changes direction; turns may be left or right, and the path may be open at the ends or "reentrant" (with ends joined). The second problem is to find a 27-block path with the maximum number of turns. In both problems the entire length of each of the eight streets must be traversed.

"When I get bored looking for *best* paths," O'Gara went on, "I like to look for *worst* ones. For example, I used to deliver mail to ten houses that were spaced at equal distances along one side of a street. What's the *longest* path a postman can take if he starts at any house, walks straight to another, then to a third and so on until he's gone once to each house?"

He made the sketch shown in Figure 166 to show how he had first tried it: from house 1 to house 6 along a path of 45 unit intervals. "And there are paths worse than that?" I asked.

O'Gara nodded. "You might ask your readers to see if they can find the worst one. If they like this kind of combinatorial puzzle, they can try the harder problem of finding a formula for the longest path as a function of *n* houses."

"Splendid," I said, scribbling in my notebook. "But I don't want to overload this interview with route problems. Have you invented any good puzzles involving other things? House numbers, for instance?"

O'Gara pulled open a drawer in which he seemed to have hundreds of problems neatly recorded on file cards. Here is one he showed me.

A long street runs east and west, with houses on both sides. Houses on one side have odd numbers in serial order, starting with 1. Houses on the other side have even numbers starting with 2. On each side there are more than 50 houses and fewer than 500. Smith lives on the odd side. The sum of all

166. A "worst-route" problem

the odd house numbers east of him exactly equals the sum of all the odd numbers west of him. The same situation holds for Jones, who lives on the even side: the house numbers west of him, on his side of the street, have the same sum as all the house numbers east of him. What are the house numbers of Smith and Jones?

"Have you ever mentioned in your department," asked O'Gara, "the old problem of the person who writes n letters, addresses n envelopes and then inserts the letters into the envelopes at random?"

"Yes," I replied, "although I gave it in terms of simultaneously dealing two decks of shuffled cards. As I recall, as n increases, the probability that no letters and envelopes will match approaches the limit of $1/e$."

"Right," said O'Gara. "With only four letters it's easy to show that the probability that one letter or more gets mailed to the right person is 5/8, and the probability that *exactly* one letter goes to the right person is 1/3."

"I'll take your word for it," I said.

"Can you tell me," he continued, smiling faintly, "the probability that exactly one of the four letters is mailed *incorrectly*?"

I started to jot down a list of all the permutations of A, B, C, D but O'Gara seized my wrist. "You have to do it in your head," he said, "and in less than 10 seconds."

I was startled for a moment, but then I broke into a laugh. Does the reader see why?

I had walked from the subway to O'Gara's apartment in a heavy downpour. When I took my leave, it was still raining. "Well," I said as we pumped hands, "you'll observe

that neither snow nor rain, nor heat, nor night can stay this courier from the swift completion of his appointed rounds."

"Ah, yes," he said, wincing. "Most everybody, I suppose, knows that statement you're paraphrasing so badly. But can you tell me who first said it?"

I could not, and I leave O'Gara's parting remark as my closing question.

Answers

A minimum-length path covering all eight streets in a square area three blocks on a side, making the minimum number of turns, can be solved with as few as ten turns, as shown in Figure 167. The solution is unique except for reflections and rotations. To prove that ten is minimal, note first that the network has eight vertices where an odd number of paths meet. According to well-known rules, the graph cannot be traversed by one continuous path (without going over any portion of the graph twice) unless the odd vertices are reduced to two or none. This can be accomplished by doubling segments of the graph, but we must do it in a way that adds as little as possible to the total length of the lines. It is easy to see that the shortest path is obtained by doubling three segments as shown in Figure 168. The doubled segments indicate portions of the original graph that must be traversed twice. This minimal path is 27 blocks long, with A and B (the two remaining odd vertices) as its end points. Many readers ignored the proviso that the *entire lengths* of all eight

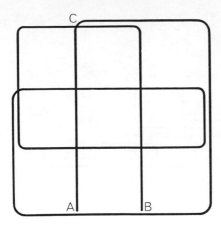

167. Minimum-turn solution

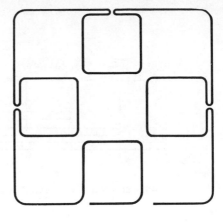

169. Maximum-turn solution

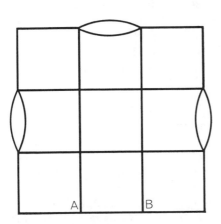

168. Minimum-turn proof

blocks must be traversed. If halves of blocks are allowed to remain untraversed, the postman can reach all his delivery spots in a minimal-length path of 23 blocks. The illustration was misleading because the problem was intended to be one of network tracing.

Five streets can be traversed their full length without a turn. If we call any segment traveled without a turn a "move," it is clear that these five streets demand at least five moves. Each of the remaining three streets requires at least two moves because each has a middle block that must be traversed twice. Therefore any continuous path from A to B must have at least 11 moves, which is the same as saying it must have at least 10 turns. Suppose we start at A and proceed to C. We cannot turn left at C because then two moves would be necessary to complete the right two-thirds of the top street, making three moves in all for this street, whereas the minimum-turn path limits this street to two. So we must turn right. Continuing in this way, analyzing all alternatives at each juncture, we find that only two travel patterns complete the trip in ten turns. One pattern is a mirror image of the other. Figure 169 shows a 27-block path with the maximum number of turns: 26. This too is unique except for reflections and rotations.

The longest path for visiting the row of

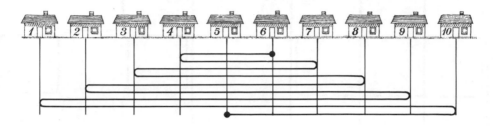

170. Answer to the "worst-path" problem

ten houses, in the second problem, is shown in Figure 170. It has a length of 49 units. When the number of houses is even, the length of the "worst" path is $\frac{1}{2}(n^2 - 2)$; when it is odd, the length is $\frac{1}{2}(n^2 - 3)$. For the derivation of both formulas see problem No. 64 in Hugo Steinhaus' *One Hundred Problems in Elementary Mathematics* (New York: Basic Books, 1964). When n is even, one end of the path must be at one of the two middle houses. When n is odd it must be at one of the three middle houses. As R. H. Shudde pointed out, the paths are not unique when n is greater than 4.

Smith's house number is 239, in a row of 169 houses. Jones's is 408, in a row of 288 houses. The solution for Smith involves finding integral solutions of $2x^2 - 1 = y^2$; for Jones, integral solutions of $2x^2 + 2x = y^2$, where x is the number of houses and y the house number. Both Diophantine equations have an infinity of solutions, but we were told that the number of houses in each case is between 50 and 500. This restricts each equation to one pair of values for x and y.

The probability that exactly one letter will go into the wrong envelope, if four are inserted at random into four envelopes, is zero, because it is impossible for three let-

ters to match their envelopes and the remaining one not to match.

The quotation, "Neither snow nor rain nor heat nor gloom of night . . . ," which is carved on the facade of New York City's General Post Office Building at Eighth Avenue and 33rd Street, is from the Greek historian Herodotus. It appears in his *History*, Book VIII, Section 98.

References

"Philately and Mathematics." Carl B. Boyer. *Scripta Mathematica*, Vol. 15, No. 2; June, 1949. Pages 105–114.

"Mathematics and Philately." H. D. Larsen. *The American Mathematical Monthly*, Vol. 60, No. 2; February, 1953. Pages 141–143.

"Mathematics on Stamps." H. D. Larsen. *The Mathematics Teacher*, Vol. 48, 1955. Pages 477–480. Vol. 49, 1956. Pages 395–396.

"The House Problem." Ralph Finkelstein. *The American Mathematical Monthly*, Vol. 72, No. 10; December, 1965. Pages 1082–1088.

"Mathematicians and Mathematics on Postage Stamps." William Schaaf. *Journal of Recreational Mathematics*, Vol. 1, No. 4; October, 1968. Pages 195–216. (For errata, see Vol. 2, No. 9; July, 1969. Page 192.)

24.   Op Art

Op (FOR "OPTICAL") topped Pop (for "popular") as the fashionable gallery art of the mid-1960's; its patterns quivered in advertisements and on dresses, bathing suits, ties, stockings, window shades, draperies, wallpaper, floor coverings, package designs, covers of math textbooks, and what have you. Op art, as everyone surely knows by now, is the name for a form of hard-edge abstractionism that has been around for half a century. Its distinguishing feature is a strong emphasis on mathematical order. Sometimes it is accompanied by effects intended to dazzle and wrench the eye: vivid colors that generate strong afterimages when the eye shifts, optical illusions, striped and dotted patterns that torture the brain like the retinal scintillations of migraine. One branch of Op art deals with moiré patterns of the type described in *Scientific American* by Gerald Oster and Yasunori Nishijima (see "Moiré Patterns," May, 1963) and by C. L. Stong ("The Amateur Scientist," November, 1964). Indeed,

Oster's shimmering patterns have been exhibited in several New York art galleries.

The Op trend, many critics say, is more than just a rebellion (like Pop) against the randomness of abstract expressionism; it reflects the growing extent to which mathematics, science and technology press on our lives. *Scientific American*, it has been observed, has been presenting Op art for years. Consider the following magazine covers: "Perfect" Rectangle, November, 1958; Reactor Fuel Elements, February, 1959; "Graeco-Latin" Square, November, 1959; "Visual Cliff" (with its distorted checkerboards, a popular Op motif), April, 1960; Spark Chamber, August, 1962; Moiré Pattern, May, 1963; and Afterimage Test Pattern, October, 1963. These covers are almost pure Op. They leave little doubt about Op's close kinship with modern science.

Although Op art is sometimes rich and warm with colors, its appeal seems to lie more in its cold, rigid, precise, unemotional and impersonal qualities. Its astonishing popularity revives ancient questions about art and mathematics. To what extent is art ruled by mathematical laws? To what extent can pure mathematical structure arouse aesthetic emotions? "The chief forms of beauty are order and symmetry and precision," wrote Aristotle in his *Metaphysics* (Book 13), "which the mathematical sciences demonstrate in a special degree." "A mathematician . . . ," declared G. H. Hardy in *A Mathematician's Apology*, "is a maker of patterns. . . . [His] patterns, like the painter's or the poet's, must be *beautiful*; the ideas, like the colors or the words, must fit together in a harmonious way. Beauty is the first test: there is no permanent place in the world for ugly mathematics."

We are surrounded on all sides, say the defenders of Op, by hard-edge squares and circles, ellipses and rectangles. The windows of a skyscraper, the streets of a city, the fronts of file cabinets, all form orthogonal patterns like a checkerboard. Why should these basic geometric designs not be reflected in our art? Opponents counter: But we want to escape from, not be reminded of, the low-order curves and 90-degree angles of a technological culture. Our eyeballs ache for random curves, impure colors and soft edges; for the patterns of leaves and clouds and water in motion. Who can write an equation for the shape of an oak tree? The mathematical structure is still there, but in nature, as in less rigid abstract art, it is more complex, more careless, and—say Op's detractors—aesthetically

less boring. (See the cover of *The New Yorker*, August 14, 1965, for Saul Steinberg's illustration of this idea.)

Whatever one's attitude toward Op, there is no denying its fascination. Nor is it surprising that many Op patterns are closely related to problems of recreational mathematics. Consider, for example, the nested and rotating squares (or rectangles) that appear in so many Op paintings and fabric designs and that whirl inward on the cover of *Scientific American*, July, 1965. The pattern can be interpreted as an illustration for the well-known "four-bug problem," which appears in Chapter 12 of my *Scientific American Book of Mathematical Puzzles and Diversions* (New York: Simon and Schuster, 1959). Four bugs at the corners of a square start to crawl clockwise (or counterclockwise) at a constant rate, each moving directly toward its neighbor. At any instant, as the bugs march toward a meeting point at the center, they mark the corners of a square, and as they crawl the square they delineate both diminishes and rotates. Each bug travels on a logarithmic spiral with a length exactly equal to the side of the original square.

If n bugs start at the corners of any regular n-sided polygon, their positions at any instant during their march will mark the corners of a similar polygon. Like the square, this polygon will shrink and turn as the bugs spiral inward. A design based on the triangular case is shown in Figure 171, originally drawn for an old issue of *Scripta Mathematica* by Rutherford Boyd. The picture contains nothing but triangles,

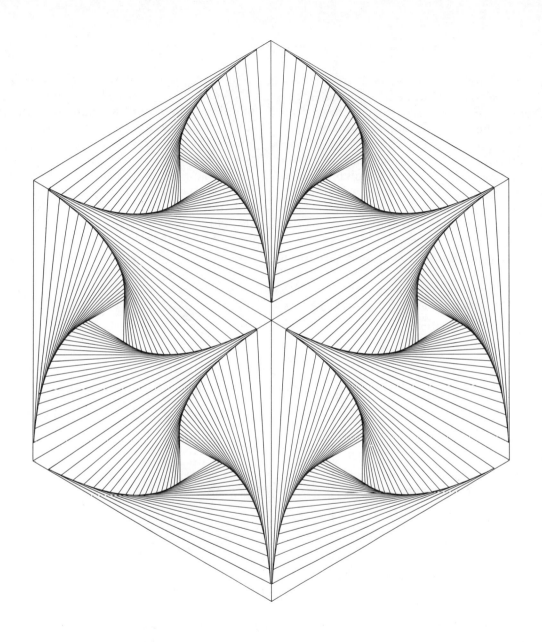

171. Design based on the "three-bug problem"

but they are hard to see because the eye is so strongly dominated by the spiral curves. In this case each logarithmic spiral is 2/3 of the original triangle's side.

For regular polygons of more than four sides the length of each bug's path is greater than a side. As J. Charles Clapham proved in the now defunct *Recreational Mathematics Magazine* (August, 1962), the length of the path of a bug starting at corner A can be found trigonometrically by extending a side AB [*see Figure 172*] and locating on it

a point X such that the angle AOX is 90 degrees. The distance AX—which is equal to r times the secant of angle θ—is the distance the bug travels. As the illustration shows, on a hexagon each bug's path is twice the length of a side.

Clapham's simple formula also applies to the square and triangular cases, and even to the degenerate "two-sided polygon"—a straight line with a zero angle θ and bugs at each end that tramp toward each other until they bump head on. At the other ex-

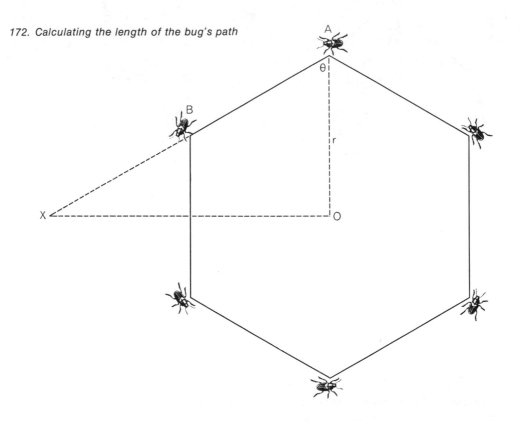

172. *Calculating the length of the bug's path*

173. Baravalle's circular "checkerboard"

treme, the circle can be considered a degenerate "infinite-sided polygon" with bugs at an infinite number of "corners." These bugs march forever around the circle like the Pine Processionary caterpillars in a famous experiment of Jean Henri Fabre's, which trailed each other for eight days around the rim of a large vase. When we apply Clapham's right triangle to the circle, sure enough, angle θ is 90 degrees and the hypotenuse is infinite.

One suspects that Op painters both here and abroad have yet to discover the thousands of eye-twisting patterns that lie buried in scientific and mathematical textbooks and back copies of academic journals. Early issues of *Scripta Mathematica*, for example, vibrate with exciting pre-Op. Figure 173 shows a striking pattern the mathematician Hermann Baravalle obtained by ruling parallel lines across concentric circles and then coloring the regions in checkerboard fashion. One might think that this pattern is topologically the same as a square checker-

174. Elliptical "checkerboard"

board—in other words, that a square checkerboard on a rubber sheet could be continuously deformed to produce the pattern. This is not the case, but it suggests a pretty puzzle. Can you cut the pattern into two parts with one straight cut so that each part is topologically equivalent to a square checkerboard? Figure 174 shows how Torbjörn Johansson, a Swedish com-

mercial artist, applied Baravalle's coloring technique to an ellipse.

In Figure 175 Baravalle has inverted every point P that lies outside the circle on the checkerboard into a corresponding point P' inside the circle, such that $OP \times OP' = r^2$, where O is the circle's center and r its radius. Every point on the plane outside the circle is thus put into one-to-one

175. *Checkerboard inversion pattern*

correspondence with every point inside. A line extending outward from the board to infinity corresponds to a line inside the central white space, extending inward toward the center but never reaching it.

Inversion geometry can, of course, be applied to three-space as easily as to the plane. An old mathematical joke says that to catch a lion you just build a cage and perform an inversion operation on the beast. The cosmos itself can be inverted and compressed inside a tennis ball. In this country during the 1870's a religious cult was actually founded on the belief that such an inverted three-space reflects the true state of affairs. Cyrus Reed Teed's "Koreshanity" put the entire universe *inside* the earth. We imagine ourselves on the outside of the earth looking out at gigantic stars scattered through an infinite space; the truth, said Teed, is that we are on the inside of a hollow earth looking *in* at small stellar bodies moving in a space that is the geometrical inverse of the space of orthodox astronomy. Teed defended his views in many books and articles; years later his ideas attracted a following in Nazi Germany. (For more details on this crazy cult see Chapter 2 of my *Fads and Fallacies in the Name of Science*; New York: Dover, 1957.)

Figure 176 shows two examples of many vertigo-inducing patterns that were studied by psychologists more than 50 years ago. They are known as "twisted-cord illusions" because they were first discovered by twisting black and white string into a single cord that was then arranged in various ways on differently patterned backgrounds. The top part of the figure consists of concentric circles (as you can prove with a compass); in the one at the bottom a spiral is made up of straight horizontal and vertical "cords" (as you can prove with a ruler).

Tesselations of the plane created by fitting together replicas of the same basic shape have long been used in design and are now turning up in many of the latest Op fabrics. The cross-pentomino appears on an Op dress advertised in 1965 by Bonwit Teller. *All* polyominoes and polyiamonds (polyiamonds are formed by joining equilateral triangles instead of squares) of order six or less will fit together to cover the plane, but so far I have seen only the cross-pentomino and the *L*-tromino (the latter on a scarf sold by Gimbels in New York City) on Op fabrics. The reader can easily create his own new Op patterns by finding ways to tile the plane with each of the 12 pentominoes and the 12 hexiamonds (for the hexiamond shapes see this book's chapter on polyiamonds).

Most, but not all, of the 108 heptominoes (for their diagrams see Solomon W. Golomb's book *Polyominoes*, pages 108–109) will tile the plane. Several British mathematicians are working on the difficult question of which of them are *not* plane-fillers. The corresponding problem for the 24 heptia-

176. "Twisted cord" concentric circles (top) and spiral (bottom)

monds [*see Figure 177*] was proposed by T. H. O'Beirne and was solved this year by Gregory J. Bishop of Boston. Only one of the 24 shapes is not a plane-filler. Can the reader identify it and prove that it cannot tesselate the plane?

The Op pattern that covers the plane with convex noncongruent heptagons [*Figure 178*] embodies a curious paradox that twiddles the brain even more than the eye. If this pattern is repeated infinitely, what is the average angle in it? Since the plane contains nothing but heptagons, and since the interior angles of any heptagon sum to 900 degrees, it follows that the average angle is 900/7 or 128⁴/₇ degrees. Note, however, that every point on the pattern is a meeting of three angles. This surely requires that the average angle be 360/3, or 120 degrees. Explain!

177. The 24 heptiamonds. Which cannot tile a plane?

178. Tesselation of convex heptagons

Answers

Were you able to make one straight cut across the circular Op pattern [*see the drawing at left in Figure 179*] so as to divide the pattern into two parts, each topologically equivalent to a square checkerboard? That the pattern itself cannot be continuously distorted to produce a checkerboard is evident from the fact that the number of its cells, 392, is not a square. Note also that the two cells inside the bull's-eye are each three-sided; any distortion that turns one of these cells into a square would turn the other into a nonconvex figure. It is therefore necessary

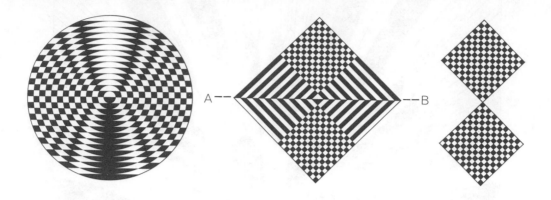

179. Answer to the topology problem

180. Proof that the V-heptiamond will not tile the plane

that the cut separate these two cells. The only straight cut that does this is one along the horizontal diameter of the large circle. The figure in the middle is topologically the same as the one at the left. It is easy to see that a single cut along *AB* produces two halves, each of which is topologically the same as a square checkerboard 14 cells on a side [*Figure 179, right*].

The only heptiamond that will not tile the plane is the V-shaped figure shown at *a* in Figure 180. The proof, by Gregory J. Bishop, an electrical engineering student at Northeastern University, is simple. A second piece can fill the colored triangular concavity of the first one only as shown at *b*. (We ignore a mirror reversal of the second piece.) The colored triangle of the second piece can now be filled only by placing a third piece as shown at *c*. The colored triangle of this figure must in turn be filled by placing a fourth piece as shown at *d*, and there is a similar lack of choice in positioning pieces 5 and 6. Now we are stuck. There is no way to fill the colored triangle associated with piece 6.

Tessellations for the other 23 heptiamonds are too numerous to illustrate. The reader may have discovered the useful trick of pairing two pieces to form a pattern that periodically tiles the plane. For example, five different heptiamonds can be paired to fit the same periodic tessellation. (Figure 181 shows four; can the reader discover the fifth?) Bishop has also established that each of the 66 octiamonds will tile the plane, and that all but four of the 108 heptominoes will do so.

181. *Tesselation for four heptiamond pairs*

The paradox of the heptagon tessellation was taken from Hugo Steinhaus' *One Hundred Problems in Elementary Mathematics*. The paradox arises from the fact that a rearrangement of terms in an infinite series can lead to a different calculation of the average term. Steinhaus gives as an example the series 1, 0, 1, 0, 1, 0 . . . for which 1/2 is the average. But the two infinite sets of ones and zeroes can also be arranged 1, 0, 1, 0, 0, 0, 0, 1, 0, 0, 0, 0, 0, 0, 0, 0, 0, 1 . . . (where successive sets of zeros have cardinal numbers that are squares of 1, 2, 3 . . .), in which case the average is 0. It is easy to form other arrangements to make the average any desired integral value between 0 and 1. In the heptagon pattern two different arrangements of two infinite sets of angles are considered, and there is no reason why the calculation of an average angle should be the same in each.

References

"Pursuit Curves and Mathematical Art." I. J. Good. *The Mathematical Gazette*, Vol. 43, No. 343; February, 1959. Pages 34–35.

"Two Hexagonal Designs." Prakash Chandra Sharma. *The Mathematical Gazette*, Vol. 45, No. 351; February, 1961. Pages 26–27.

"Op Art: Pictures That Attack the Eye." *Time*, Vol. 84, No. 17; October 23, 1964. Pages 78–84.

"Op Art." *Life*, Vol. 57, No. 24; December 11, 1964. Pages 132–140.

"Art That Pulses, Quivers and Fascinates." John Canaday. *The New York Times Magazine*, Vol. 114, No. 39,110; February 21, 1965. Pages 12–59.

"Optical Art." Gerald Oster. *Applied Optics*, Vol. 4, No. 11; November, 1965. Pages 1359–1369.

Optical Illusions and the Visual Arts. Ronald G. Carraher and Jacqueline B. Thurston. New York: Reinhold, 1966.

Cybernetic Serendipity: The Computer and the Arts. Edited by Jasia Reichardt. London: *Studio International*, 1968. (Special issue.)

25. Extraterrestrial Communication

Across the gulf of space, minds that are to our minds as ours are to those of the beasts that perish, intellects vast and cool and unsympathetic, regarded this earth with envious eyes, and slowly and surely drew their plans against us.

H. G. Wells, *The War of the Worlds*

IN 1898, WHEN WELLS'S NOVEL was first published, a number of distinguished astronomers seriously believed Mars was inhabited by creatures with "intellects vast and cool" and superior to our own. The Italian astronomer Giovanni Schiaparelli (the uncle of the dress designer Elsa Schiaparelli) had reported in 1877 that he saw fine lines crisscrossing the red planet. A wealthy Bostonian, Percival Lowell, became so excited by Schiaparelli's continued disclosures that he decided to abandon Oriental studies and become an astronomer. In 1894, when Mars was unusually close to the earth, Lowell established his own observatory on "Mars Hill" in Flagstaff, Arizona.

Lowell too saw the lines Schiaparelli had called "canali." (The word, which means "channels," had been subtly mistranslated "canals.") Indeed, he saw them in fantastic profusion; eventually he mapped more than 500. In three books — *Mars* (1895), *Mars and Its Canals* (1906) and *Mars as the Abode of Life* (1908) — Lowell argued that the lines he saw were wide bands of vegetation bordering enormous irrigation ditches constructed to bring water from melting polar caps to the dry Martian deserts. "That Mars is inhabited by beings of some sort or other," he wrote, "we may consider as certain as it is uncertain what those forms may be." Lowell's Mars books had an enormous influence on early science fiction; the canals turned up everywhere, from Wells's 1897 short story "The

Crystal Egg" to the later Martian romances of Edgar Rice Burroughs.

There is no doubt about Lowell's competence as an astronomer. Calculations he made in 1915 led to the discovery of Pluto by Clyde W. Tombaugh in 1930—at the Lowell Observatory. ("Pluto" was chosen as the planet's name because its first two letters are Lowell's initials and its last two the beginning letters of Tombaugh; Pluto's symbol, ♇, combines *P* and *L*.) But as Jonathan Norton Leonard observes in his book *Flight into Space*, Lowell's temperament was closer to that of his sister Amy, the cigar-smoking poet, than to that of his cautious, conservative brother Abbott Lawrence, who became president of Harvard. Although some astronomers enthusiastically confirmed Lowell's observations of Martian canals, others with better telescopes and better eyes could see no canals at all. Even in today's best telescopes Mars is a tiny, jiggling spot of light, and in those rare, fleeting moments when the image holds still, one's mind can play strange tricks. Photographs are no help because the earth's turbulent atmosphere blurs the image.

The consensus among astronomers today is that Schiaparelli, Lowell and their followers were the victims of optical illusions induced by irregular splotches on the red planet and elaborated by astigmatism and psychological self-deception.

Among the few living scientists who continue to take Lowell's speculations seriously the most vocal is Wells Alan Webb, a California chemist. In his book *Mars, the New Frontier: Lowell's Hypothesis* (1956) and in many magazine articles and lectures he has reported on an interesting topological analysis of canal drawings made by Lowell and by one of his leading supporters, Robert J. Trumpler. Considering the maps of these two astronomers as geometrical networks, Webb determined the percentage of vertices at which three, four, five, six, seven, and eight rays came together. On the maps drawn by both men vertices of order 4 (four lines meeting at a point) predominate: they constitute about 43 percent of the vertices on Trumpler's maps and about 55 percent on Lowell's. A similar analysis of networks found in nature—mud cracks, shrinkage cracks of glazed chinaware, cracks in ancient lava beds, rivers, and so on—showed order-3 vertices leading their percentage list. Only in networks constructed by living things, such as spider webs and animal trails, did Webb find the order-4 points predominating. The networks that are topologically most like the Lowell-Trumpler maps are such man-made ones as railroad lines and air travel routes. Thus does topology, Webb argues, back up Lowell's intuitive conviction that the canals must have been the work of high-order intellects.

Webb's arguments assume, of course, a correspondence between the Martian surface and the Lowell-Trumpler maps. But if these maps are no more than doodles of what Lowell and Trumpler imagined they saw, their topological similarity to railroad lines is easily understood. At this writing the first television pictures of Mars are being received from *Mariner IV*, but they have not yet established whether or not there are lines on the planet. Certainly few astronomers expect the pictures to show anything like Lowell's cobwebs; if they do, the great canal controversy will surely break out again.

From 1880 to 1925, when interest in Martian canals was high, all sorts of proposals were put forward for establishing contact with Martians. Two frequent suggestions were that a powerful searchlight be built that would blink a code message, or that a chain of bright lights be stretched across a vast area to make a diagram, visible in Martian telescopes, of the Pythagorean theorem. There was much discussion about radio contacts: sending a series of beeps to represent the counting numbers (beep; beep, beep; beep, beep, beep; . . .) or such arithmetical trivia as two plus two equals four. In 1900 Nikola Tesla declared that he had received radio signals from intelligent beings on Mars. Twenty-one years later Guglielmo Marconi made a similar announcement. Spiritualists too were in frequent contact with minds on the red planet. The most remarkable was Hélène Smith, a Swiss medium, whose strange story is told in the book *From India to the Planet Mars: A Study of a Case of Somnambulism with Glossolalia* (1900) by the Swiss psychologist Théodore Flournoy. In her trances Hélène seemed to be under Martian control, speaking and writing a complex Martian language, complete with its own alphabet. (On Hélène, and other mediums who claimed Martian contacts, see also

Chapter 8, "From Kensington to the Planet Mars," in Harry Price's *Confessions of a Ghost-Hunter*; New York: Putnam, 1936.)

Now that we are close to landing exploratory robots on Mars and are expecting to find, at the most, only a low-grade vegetation, interest in extraterrestrial communication has shifted to planets in other solar systems. In 1960 Project Ozma failed to detect any radio messages from outer space after several months of listening near the frequency at which free hydrogen radiates. (For various reasons this frequency, with its wavelength of 21 centimeters, seems to be the ideal frequency for interstellar communication.) Nevertheless, interest both in sending and in searching for such messages continues, and much abstruse work is being done on the best methods of exchanging information with an alien culture once contact is established. It is a fascinating problem, almost the exact opposite of devising wartime codes. The purpose of a code is to transmit information in such a way as to make it as difficult as possible for anyone not knowing the key to understand the message. The purpose of an interstellar code is to communicate with minds that know nothing of our language, and in such a way as to make it as *easy* as possible for them to understand.

Many of the papers in *Interstellar Communication*, edited by A. G. W. Cameron (1963), are concerned with this task. All experts agree that messages had best start with simple arithmetic. One assumes that units can be counted by any type of intelligent creature, and that arithmetical laws are uniform throughout the galaxy. Of course one cannot assume that any given method of symbolizing numbers — such as our positional notation based on 10 — would be universal. It would be foolish, for example, to try to get extraterrestrial attention by transmitting a decimal expansion of pi; the aliens might use a different base system and our pi would seem at first to be no more than a series of random symbols. Hans Freudenthal, a Dutch mathematician, has invented an elaborate artificial language he calls Lincos (for "lingua cosmica") that starts out with arithmetic and simple logic, proceeds to more advanced mathematics and ultimately is capable of communicating all human knowledge. The first volume of his work, *Lincos: Design of a Language for Cosmic Intercourse*, was published in the Netherlands in 1960.

Most of Freudenthal's efforts may prove to be irrelevant because of the great ease with which pictures can be sent by a simple code of two symbols. This does not require that beings receiving such a code have eyes sensitive to light but only that they have some means of mapping the shape of things; our visual pictures could be translated by them into whatever sensory technique provides *their* best way of observing the world. Perhaps the simplest way to transmit a shape is by a two-symbol message giving directions for scanning a rectangular matrix of cells, one symbol indicating that a cell is filled and the other that it is empty. Indeed, this is the technique by which pictures are now transmitted by radio as well as the basis of television-screen scanning.

Consider the following 100-symbol message.

```
0000000111
1111111101
1110000111
1010000000
0000000000
1010110101
1010100101
1100110111
1010100010
1010110010
```

The 100 symbols suggest the 10-by-10 matrix shown in Figure 182. If the reader will scan the cells from left to right, top to bottom, darkening every cell indicated by 1, he will produce a picture of a familiar object and the English word for it. It is easy to see that once the principle of picture scanning is grasped, ease in communication advances by leaps and bounds.

Since it might take hundreds or thousands of years for a message to travel from the earth to a planet in another solar family, it obviously is impossible to chat back and forth the way one does on a telephone. Messages would have to open with something designed to catch attention — the counting numbers or a series of primes — followed by simple arithmetic leading quickly to picture scanning, then on to encyclopedic transfers of information. But what sort of information should be sent first? Here we come up against a curious situation. One might suppose that the simplest knowledge to send would be about things physicists call "observables" — information

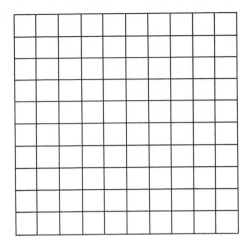

182. 10-by-10 scanning matrix

derived from our senses, often aided by relatively simple observational devices such as telescopes and microscopes. But suppose the minds on Planet X have as their most highly developed sense some method of mapping the world that evolution here has failed to exploit, say by magnetic forces or some type of radiation not yet known to us. Our pictures of the world, derived from our observables, might have less meaning on Planet X than information about such "unobservables" as electrons, protons and neutrons. If so, the inhabitants of Planet X might understand a description of the periodic table of elements more readily than a description of a house or tree. From one point of view the colors, shapes and sounds of our world are the bedrock facts and the electron a shadowy abstraction. The problems arising here suggest the opposite. The mathematical structure of a helium atom

may be more universally understood than the color, smell, taste and shape of an apple, not just because apples are unlikely to grow on other planets but because other minds may map their worlds with senses that have little in common with sight, smell, taste and touch. Inferred entities such as particles and electromagnetic fields might be easier for extraterrestrials to understand than the familiar sights and sounds of our world.

In 1960 Ivan Bell, an Englishman teaching English in Tokyo, read about the plans for Project Ozma. To amuse his friends he devised a simple interplanetary message of 24 symbols. It was printed in *The Japan Times* of January 22, 1960, and readers were asked if they could decipher it. Four complete solutions were received. One was from Mrs. Richard T. Field, now living in Bridgeton, New Jersey, who sent me a photocopy of Bell's article.

Bell's message is reproduced in Figure 183. It is much easier to decipher than it looks, and readers are urged to try it. Letters from *A* through *Z* (omitting *O* and *X*) provide the 24 symbols. (Each symbol is presumably radioed by a combination of beeps, but we need not be concerned with those details.) The punctuation marks are not part of the message but indications of time lapses. Adjacent letters are sent with short pauses between them. A space between letters means a longer pause. Commas, semicolons, and periods represent progressively longer pauses. The longest time lapses come between paragraphs, which are numbered for the reader's convenience; the numbers are not part of the

1. A. B. C. D. E. F. G. H. I. J. K. L. M. N. P. Q. R. S. T. U. V. W. Y. Z.

2. A A, B; A A A, C; A A A A, D; A A A A A, E; A A A A A A, F; A A A A A A A, G;
    A A A A A A A A, H; A A A A A A A A A, I; A A A A A A A A A A, J.

3. A K A L B; A K A K A L C; A K A K A K A L D. A K A L B; B K A L C; C K A L D;
    D K A L E. B K E L G; G L E K B. F K D L J; J L F K D.

4. C M A L B; D M A L C; I M G L B.

5. C K N L C; H K N L H. D M D L N; E M E L N.

6. J L A N; J K A L A A; J K B L A B; A A K A L A B. J K J L B N; J K J K J L C N.
    F N K G L FG.

7. B P C L F; E P B L J; F P J L F N.

8. F Q B L C; J Q B L E; F N Q F L J.

9. C R B L I; B R E L C B.

10. J P J L J R B L S L A N N; J P J P J L J R C L T L A N N N. J P S L T; J P T L J R D.

11. A Q J L U; U Q J L A Q S L V.

12. U L W A; U P B L W B; A W D M A L W D L D P U. V L W N A; V P C L W N C.
    V Q J L W N N A; V Q S L W N N N A. J P EWFGH L EFWGH; S P EWFGH L EFGWH.

13. GIWIH Y HN; T K C Y T. Z Y CWADAF.

14. D P Z P W N N I B R C Q C.

183. Ivan Bell's interplanetary message

message. To minds in any solar system the message should be crystal clear except for the last paragraph, which is somewhat ambiguous; even if properly deciphered, it could be understood fully only by inhabitants of one of our solar system's planets. (Mrs. Field wrote her answer in the same code and signed off by saying—in the code—that she lived on Jupiter.) The key to Bell's message and a complete translation will be found in the answer section.

Addendum

The following exchange, prompted by my writing about the foregoing topics, was printed in the Letters column of *Scientific American*, January, 1966:

Sirs:

I strongly disagree with some statements made by Martin Gardner in the "Mathematical Games" section of your August [1965] issue.

It is quite fashionable to relegate the Martian canals to illusions. The people who subscribe to such a view have little or no experience in actually looking at planetary detail. There are several observing parameters necessary to see the delicate subdivisions in Saturn's rings, the intricate "festoons" on Jupiter, canals and oases on Mars, and minute clefts and craters on the moon.

Gardner writes: "Although some astronomers enthusiastically confirmed [Percival] Lowell's observations of Martian canals, others with better telescopes and better eyes could see no canals at all." The latter part of this statement is incorrect. The Hartmann tests and the optical performance of the 24-inch Lowell and 36-inch Lick refractors show that these telescopes are of extraordinary optical quality. I have spent several hundred hours observing the moon and the planets with the Lowell 24-inch refractor. Only during the rarest moments of freedom from our own atmospheric turbulence are the finest recorded details visible. I would estimate that such quality seeing exists for less than a hundredth of one percent of the time one spends looking through the eyepiece. Now, what are the chances of an astronomer's seeing very fine planetary detail if he takes a look at a planet for a few minutes before beginning his stellar program of observing for the night?

The late Dr. Robert Trumpler of the Lick Observatory was doubtful of the canals on Mars in his younger days. Then he decided to observe Mars carefully throughout the 1924 opposition. He saw and recorded a great number of canals, as attested by his maps of Mars. Dr. E. C. Slipher also had keen eyes and saw well.

You learn to see fine planetary detail by much looking through a good telescope, not by sitting in an armchair in a warm office. A person with normal eyesight can see canals when the seeing is extraordinarily good, provided that the optical parameters of the telescope are proper. Anyone who uses a large aperture with too low a magnifying power obtains a dazzlingly brilliant image. This drowns out delicate dark detail because of the fierce irradiation from the over-brilliant surrounding area on the planet's disk.

In all probability the canals are likely to be discontinuous dark patches in an alignment. My studies of Mars indicate that the canals are crustal faults and are discontinuous where dust has covered up portions of them. The boundaries of the angular maria are exactly aligned with canals that proceed across the Martian deserts for hundreds of miles to dark spots called "oases," which are probably large impact craters (see

The Astronomical Journal, October, 1950, page 184).

There are certainly portions of canals visible on some of the *Mariner IV* pictures. You cannot measure truth by popular vote, because only a few learn the difficult techniques of seeing fine planetary detail from much patience and experience. The stronger canals of Mars are also recorded on photographs. One does not solve a scientific phenomenon by wishing it out of existence.

CLYDE W. TOMBAUGH

Observatory
New Mexico State University
University Park, New Mexico

Sirs:

The controversy among astronomers, between the small minority who report seeing straight lines on Mars and the majority who have been unable to see them, has been a bitter one that is not yet laid to rest. "The only possible explanation of the differences," wrote the eminent British astronomer H. Spencer Jones in his *Life on Other Worlds*, "is that the observation of these faint elusive details is subject to complex personal differences. . . . Subconscious interpretation of what is faintly glimpsed may be very different for two different persons. The eye of one may tend to bridge the gap between faint details and to draw a marking as a uniform, straight, continuous line unless he can clearly see that there are irregularities, bends and discontinuities in it. Another may only draw it in this way when he can see beyond the possibility of doubt that it is uniform, straight and continuous." Jones cites an experiment in which dots, shady patches and short lines were randomly drawn on a sheet of paper, and a class of children was asked to draw what it saw. Many of the children, particularly those in back seats, connected the prominent features with straight lines.

One of the strongest indications that Giovanni Schiaparelli and Percival Lowell, the first two astronomers to map the Martian "canals," were victims of optical illusions is that both men reported an unaccountable "doubling" of canals. Over a period of days, or even hours, certain canals were mysteriously and temporarily transformed into two parallel lines. Lowell reported seeing hundreds of such instances, although it was pointed out at the time that the distances on Mars, between pairs of parallel lines, were much too small for the resolving power of the lenses he was using.

Dr. Tombaugh's more moderate view, that there are linear structures on Mars and that they are crustal faults connecting impact craters, is similar to one advanced by Alfred Russel Wallace in a fascinating and perhaps prophetic little book called *Is Mars Inhabitable?* (London, 1907). Wallace disagreed with Lowell's belief that the canals were the work of intelligent beings—indeed, he concluded that Mars was not only uninhabited but also "absolutely uninhabitable"—although he did not question the existence of a canal network. He argued that Mars had been so heavily pelted by meteors that its surface became molten. As the planet cooled, meteors continued to fall, causing more craters that became weak spots in the crust. As the crust continued to cool and shrink, it cracked along straight lines that joined these large impact craters.

In the next few years we may learn exactly to what extent linear features exist on the Martian surface, although the controversy could drone on as a cloudy semantic quarrel over whether certain features should be called "linear." In my opinion the word "canal" should be reserved for the long, sharply defined, extremely straight threadlike lines mapped by Schiaparelli, Lowell, Trumpler and others and not applied to the broad, hazy, irregular markings that show on some photographs.

MARTIN GARDNER

Answers

The key to the 24 symbols in Ivan Bell's interplanetary message is shown in Figure 184. The message translates as follows:

1. [This simply states the 24 symbols.]

2. [This identifies the first 10 symbols (A through J) with the numbers 1 through 10.]

3. [Symbols for "plus" and "equals" are introduced.] $1 + 1 = 2$; $1 + 1 + 1 = 3$; $1 + 1 + 1 + 1 = 4$. $1 + 1 = 2$; $2 + 1 = 3$; $3 + 1 = 4$; $4 + 1 = 5$. $2 + 5 = 7$; $7 = 5 + 2$. $6 + 4 = 10$; $10 = 6 + 4$.

4. [The minus sign is introduced.] $3 - 1 = 2$; $4 - 1 = 3$; $9 - 7 = 2$.

5. [Zero is introduced.] $3 + 0 = 3$; $8 + 0 = 8$. $4 - 4 = 0$; $5 - 5 = 0$.

6. [Positional notation, based on 10, is introduced. $J = AN$ translates J into the decimal form 10.] $10 + 1 = 11$; $10 + 2 = 12$; $11 + 1 = 12$. $10 + 10 = 20$; $10 + 10 + 10 = 30$. $60 + 7 = 67$.

7. [The multiplication symbol is introduced.] $2 \times 3 = 6$; $5 \times 2 = 10$; $6 \times 10 = 60$.

8. [The division symbol is introduced.] $6 \div 2 = 3$; $10 \div 2 = 5$; $60 \div 6 = 10$.

9. [Exponents are introduced.] $3^2 = 9$; $2^5 = 32$.

A	1
B	2
C	3
D	4
E	5
F	6
G	7
H	8
I	9
J	10
K	$+$
L	$=$
M	$-$
N	0
P	$\times$
Q	$\div$
R	to the power of
S	100
T	1,000
U	1/10
V	1/100
W	. [decimal point]
Y	$\cong$ [is approximately equal to]
Z	π [pi]

184. Key to the interplanetary message

10. [Symbols for 100 and 1,000 are introduced.] $10 \times 10 = 10^2 = S = 100$; $10 \times 10 \times 10 = 10^3 = T = 1,000$. $10 \times 100 = 1,000$; $10 \times 1,000 = 10^4$.

11. [Symbols for 1/10 and 1/100 are introduced.] $1 \div 10 = 1/10$; $1/10 \div 10 = 1 \div 100 = 1/100$.

12. [The decimal sign is introduced.] $1/10 = .1$; $1/10 \times 2 = .2$; $1.4 - 1 = .4 = 4 \times 1/10$. $1/100 = .01$; $1/100 \times 3 = .03$. $1/100 \div 10 = .001$; $1/100 \div 100 = .0001$. $10 \times 5.678 = 56.78$; $100 \times 5.678 = 567.8$.

13. [The sign for "approximately equal to" is introduced.] $79.98 \cong 80$; $1,000 + 3 \cong 1,000$. [The sign for pi is introduced.] $\pi \cong 3.1416$.

14. $\dfrac{4 \times \pi \times .0092^3}{3}$.

The final statement is the formula for the volume of a sphere with a radius of .0092. As Bell recognized when he gave the answer to his message (*Japan Times*, January 29, 1960), there is an ambiguity here that could have been avoided only if information about the use of brackets or the order in which arithmetical operations are to be performed had been given previously. The formula suggests that an actual sphere is being described. If the receivers of the message are on a planet in our solar system, they should be clever enough to deduce that the sun's radius is providing the unit of length, and that the radius of the third planet from the sun is .0092 of the sun's radius. The expression therefore gives the volume of the earth and is a sign-off statement indicating the source of the message.

References

Lincos: Design of a Language for Cosmic Intercourse. Hans Freudenthal. New York: Humanities (North-Holland), 1960.

"Extraterrestrial linguistics." Soloman W. Golomb. *Astronautics*, Vol. 6, No. 5; May, 1961. Pages 46–47.

Interstellar Communication. A. G. W. Cameron, Ed. New York: W. A. Benjamin, Inc., 1963.

We Are Not Alone. Walter Sullivan. McGraw-Hill Book Company, 1964.

Intelligent Life in the Universe. I. S. Shklovskii and Carl Sagan. San Francisco: Holden-Day, 1966.

"Is Anyone There?" Isaac Asimov. *Is Anyone There?* New York: Doubleday, 1967. Chapter 22.